Aerobic Granular Sludge

WATER AND ENVIRONMENTAL MANAGEMENT SERIES

Aerobic Granular Sludge

Selected proceedings of the 1st IWA-workshop Aerobic Granular Sludge organised by the Institute of Water Quality Control and Waste Management of the Technical University of Munich (TUM) in cooperation with the Institute of Advanced Studies on Sustainability of the European Academy of Sciences and Arts (EASA) and the International Water Association (IWA)

Edited by S. Bathe, M.K. de Kreuk, B.S. Mc Swain and N.Schwarzenbeck

Published by IWA Publishing, Alliance House, 12 Caxton Street, London SW1H 0QS, UK
Telephone: +44 (0) 20 7654 5500; Fax: +44 (0) 20 7654 5555; Email: publications@iwap.co.uk
Web: www.iwapublishing.com

First published 2005

ISBN: 978-1-84339-509-6
ISSN: 1476-1785

IWA Publishing

Contents

Preface

Aerobic Granular Sludge has recently received growing attention by researchers and technology developers, worlswide. Laboratory studies and preliminary field test have been executed leading to the conclusion that granular activated sludge can be readily established and profitably used in activated sludge plants, provided "correct" process conditions are chosen. But what makes process conditions "correct"? And what makes granules different from activated sludge flocs? Answers to these questions offered by the various researchers were quite different, and they have changed as knowledge about the granulation processes evolved. Time appeared to be ripe to bring together the core group of researchers and technology developers working on activated sludge granulation, discuss the various aspects of granulation and granula application, exchange knowledge, develop a common nomenclature, and explore open research questions. Consequently, the International Water Association (IWA) in cooperation with the Institute of Water Quality Control and Waste Management of the Technical University of Munich (TUM) and the Institute of Advanced Studies on Sustainability of the European Academy of Sciences and Arts (EASA) organized an international workshop which was held on September 27 and 28, 2004 in Munich, Germany.

The papers presented during the workshop and a summary of the discussions are compiled in this issue of the IWA Water and Environment Management Series. Invited were papers pertaining to the formation, structure, and application of aerobic granular sludge. The major topics for the workshop included:

- Reasons and mechanism of aerobic granule formation
- Structure of the microbial population of aerobic granules
- Role, composition and physical properties of extracellular polymeric substances
- Diffusion limitation and microbial activity within granules
- Physico-chemical characteristics
- Operation and application of granule reactors
- Scale-up aspects of granular sludge reactors, and case studies

The papers which were received prior to the meeting underwent a rigorous peer review process. 18 papers were eventually selected, but the time allowed for oral presentation was limited to only 10 minutes to leave ample time for discussion. In fact, seven hours of the discussion time were allocated and actively filled by the delegates with top-level discussion.

The participants attending the workshop represented countries such as China, France, Germany, Italy, Japan, Mexico, Singapore, Spain, South Africa, The Netherlands, and Turkey. Most of them had never been to the Munich Octoberfest which happened to take place during the time the workshop was held. Happily, everybody joined in an excursion to the "Wies´n", participated in table dancing and singing, and everybody managed to drink his or her share of Octoberfest beer, be it a litre or two. No headaches the next day, but a great deal of enhanced openness to exchange ideas and visions. Excellence in research depends not only on bright brains but on friendship and trust among scholars as well. Small IWA specialised conferences and workshops proved to be so successful because they provided a unique opportunity to get to know each others scientific excellence as well as each others personality. It is certainly worth to keep cultivating the IWA Specialised Conference culture.

Munich, 23 – October - 2004 Peter A. Wilderer

The Unity of Biofilm Structures

M.C.M. van Loosdrecht, M.K. de Kreuk and J.J. Heijnen

Department of Biotechnology, Delft University of Technology, Julianalaan 67, 2628BC, Delft, The Netherlands (E-mail: m.dekreuk@tnw.tudelft.nl; m.c.m.vanloosdrecht@tnw.tudelft.nl)

Abstract The morphology of biomass structures has been studied in many different ways for the last decades. However, the complexity of biofilm or granular structures makes it difficult to find cause and effect in questions concerning morphogenesis. With all stated research questions, one can ask if it is the microbiology or the process conditions that determines the structure of the aggregates. In this paper, biofilm, aerobic granular sludge, and bulking sludge experiments, as well as biofilm modeling studies, are presented to show that process conditions are of major importance to determine the final biofilm structure.

Keywords: Aerobic granular sludge, biofilm, growth rate, process parameters, shear

Introduction

Over the past ten years, much research has been performed in the field of aerobic granular sludge. This research has two reasons: first studies are performed because of a general interest in background and morphogenesis of bacterial structures. Secondly, granular sludge has the potential to form the base of a new compact wastewater treatment system.

Morphology of biomass structures is studied in many different ways: syntrophic relations of bacteria in anaerobic granular sludge studies, presence of specific organisms in bulking sludge, and the role of the microorganisms in building the porous structure of biofilms. Behind all these research questions, one can ask if it is the microbiology or the process conditions that are decisive for structuring the morphology.

In general biofilm structures are highly complex ecosystems, according to microbiology, morphology and flow. In these systems, the structure of the biofilm influences the activity (position of different organisms, porosity), and the activity of the organisms influences the structure of the biofilm. Due to this complexity of biofilm structures, it is very difficult to find cause and effect in questions concerning the morphogenesis. It is difficult to get representative samples from most biofilm processes. For example, how many granules should be analyzed by a microelectrode or sliced for FISH analysis before a statistically significant conclusion can be drawn about the average position of different groups of organisms in a structure? It is also very difficult to change only one experimental parameter in biofilm experiments. As an example, changing a concentration causes changed fluxes, penetration depth, biological activity and eventually granule structure. In this case, modeling can be used in order to evaluate efficiently the relative importance of parameters on the biofilm or granule system.

Biofilm Complexity

From experiments in existing biofilm systems as expended granular sludge bed (EGSB) reactors, Internal Circulation (IC) reactors and Biofilm airlift systems, it can be seen that three different factors lead to the same change. An increased surface loading rate, a decreased shear force and an increased potential growth rate of the microbial population all lead to thicker, less dense and more porous biofilms/granules. Additionally, it has been shown in steady-state systems that a large part of detaching biomass is growth related, not shear related as generally assumed (Van Loosdrecht *et al.*, 1995). Experiments also showed that the redox gradient forms a more important parameter in granule structure than the growth rate gradient. In other words, biofilm structures are primarily affected by the concentration of the electron acceptors, and secondly by the maximum growth rate of the specific microbial populations. For example, anaerobic organisms can only grow in the anaerobic core of a granule (lowest redox), while their actual growth rate can be higher than the nitrifying organisms that are more on the outside in the granule at high redox potentials.

The general hypothesis of the formation of smooth and dense biofilms that was postulated is: the lower the actual growth rate and the higher the shear stress, the smoother the biofilms will become (Van Loosdrecht *et al.,* 1995, 1997) (Figure 1).

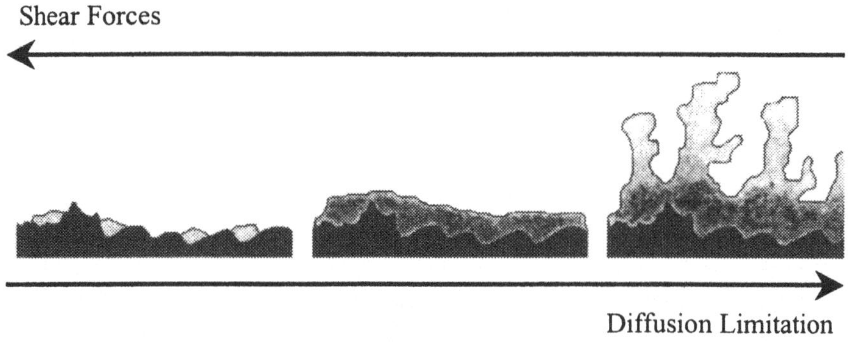

Figure 1: Schematic representation of the interaction between growth and detachment on the biofilm (or granule) structure (From: Van Loosdrecht *et al.*, 1995)

Lessons from Modelling

As mentioned above, biofilm models can help to understand the importance of different parameters. In the 1970's, biofilm modeling started with a 1- dimensional model, in which a uniform biomass was defined with a substrate gradient in the biofilm. This model was extended to a multi-species and multi-substrate biofilm model, which was still 1D. In the late 1990's, 2-D and even 3-D multi-species and multi-substrate biofilm models were developed, in which the change of substrate concentration, growth rate and biomass density and behavior is determined for every biofilm domain in the modeled reactor. In this model, different relations for the main processes are solved:

1) Laminar flow equations according to Navier-Stokes relations;
2) Diffusion, convection and reaction in mass balances of all available substrates;
3) Mass balances of the biomass, growth;
4) Discrete or continuous biomass spreading, in which biomass shifts to a different position when the actual particle concentration exceeds the maximum particle concentration in a certain domain;
5) Mechanical equilibrium equations determining the stresses and strains on the biofilm;
6) Discrete and continuous biofilm detachment.

Different models based on different calculatory procedures (cellular automata, continuum, and individual particles) were developed, all leading to a good description of the experimental observations that a fingertype outgrowth of the biofilm forms at increased substrate gradients in the biofilm or increased substrate limitation (Figure 2). The main conclusion that can be drawn from biofilm modeling results, so far, is that the morphology of the biofilms are a result of interaction between growth velocity of the biofilm front, depending on growth rate and loading rate, and the shear rate. Furthermore, there is no need to assume specific microbiological processes influencing biofilm morphology in order to explain the present observations of biofilm systems.

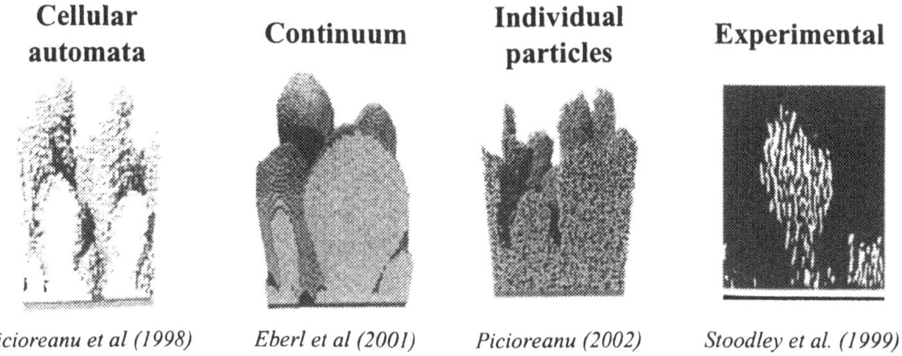

Figure 2: Different models based on three different calculatory procedures that lead to the same description of experimental results

Applications to Granular Sludge

Lessons from biofilm experiments and models show that there are different circumstances under which granule formation can be expected. First, sufficient shear is needed in relation to the maximal growth rate of the type of organisms (methanogens versus acidogens, nitrifiers versus heterotrophs; the latter groups will form less dense structures under the same circumstances as the first groups and would require higher shear rates). Second, high substrate concentrations are needed, in order to prevent strong gradients inside the granule. This can be obtained in a plug-flow or SBR process. Third, selection for slow growth is necessary (as obtained by applying a feast-famine regime if needed). For aerobic granule formation, this last condition seems to be crucial. This can be met by the conversion of readily degradable COD in slowly degradable substances, as internal storage polymers. To convert all available easy degradable COD, phosphate or glycogen accumulating organisms should be selected. Research showed that this leads so far to the most stable aerobic granules under unfavorable circumstances at low dissolved oxygen concentrations and low shear stress (De Kreuk and Van Loosdrecht, 2004).

One question that rises from the aerobic granular sludge research is if bulking sludge is the other side of the same medal. In other words, are the same mechanisms responsible for the existence of filamentous bulking sludge flocs as the mechanisms responsible for aerobic granule formation? From bulking sludge research it was found that the specific maximum substrate uptake rates in bulking sludge are more or less equal to the specific maximum substrate uptake rates of well settling sludge ($-q_s^{max}{}_{\text{bulking sludge}} \approx -q_s^{max}{}_{\text{well settling sludge}}$).

Furthermore, the specific PHB production rate of the two types of sludge is also more or less similar ($q_p^{max}{}_{bulking\ sludge} \approx q_p^{max}{}_{well\ settling\ sludge}$) (Martins *et al.*, 2004). These two observations contradict the conventional kinetic selection theory and underlines that the substrate concentration in bulking sludge phenomena is important as well. Experiments show that with decreasing substrate concentration and therefore decreasing substrate uptake rates, activated sludge flocs become more and more irregular. When the substrate uptake rate is bigger than 0.8 times the maximum substrate uptake rate granules with a diameter larger than 0.5 mm will form and that only a few filaments exist. With substrate uptake rates between 0.6 and 0.8 times the maximum uptake rate, fingertype structures come into existence and flocs become more porous, while still few filaments are present. Substrate uptake rates below 0.6 times the maximum uptake rate lead to filamentous outgrowth and fingertype structures (Figure 3) (Martins *et al.*, 2003).

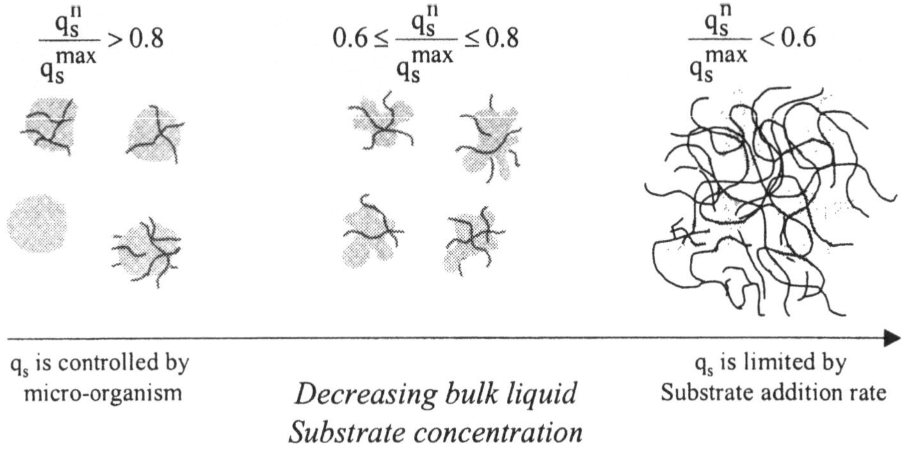

Figure 3: Schematic representation of the morphology of the sludge at different substrate uptake rates under aerobic conditions without dissolved oxygen limitation (from: Martins *et al.*, 2003)

Unity of Biofilm Structures

Experiments and modeling of bulking sludge, granule formation and biofilms led to the same conclusions towards structure and morphology. It can be said that these different morphologies of microbial structures are outings of the same phenomena and can be explained from the same basic principles. In conclusion, it can be stated that there is no need for theories based on specific microbial processes and that process conditions are of major importance for the final structure.

References

De Kreuk, M. K. and Van Loosdrecht, M.C.M. (2004) Selection of Slow Growing Organisms as a Means for Improving Aerobic Granular Sludge Stability. *Wat. Sci. Technol.* **49**(11-12), 9-19.

Eberl, H. J., Parker, D.F. and Van Loosdrecht, M.C.M. (2001) A new deterministic spatiotemporal continuum model for biofilm development. *J. Theor. Med.* **3,** 161–175.

Martins, A. M. P., Heijnen, J.J. and Van Loosdrecht, M.C.M. (2003) Effect of feeding pattern and storage on the sludge settleability under aerobic conditions. *Wat. Res.* **37**(11), 2555-2570.

Martins, A. M. P., Pagilla, K., Heijnen, J.J. and Van Loosdrecht, M.C.M. (2004) Filamentous bulking sludge - a critical review. *Wat. Res.* **38**(4), 793-817.

Picioreanu, C., Kreft, J. U. and Van Loosdrecht, M.C.M. (2004) Particle-based multidimensional multispecies Biofilm model. *Appl. Env. Micrbiol.* **70**(5), 3024-3040.

Picioreanu, C., Van Loosdrecht, M.C.M. and Heijnen, J.J. (1998). Mathematical modeling of biofilm structure with a hybrid differential-discrete cellular automaton approach. *Biotechnol. Bioeng.* **58**(1), 101-116.

Stoodley P., Boyle, J.D., De Beer, D. and Lappin-Scott, H.M. (1999) Evolving perspectives of biofilm structure. *Biofouling* **14**(1), 75–90.

Van Loosdrecht, M. C. M., Eikelboom, D. H., Gjaltema, A., Mulder, A., Tijhuis, L. and Heijnen, J.J. (1995) Biofilm structures. *Wat. Sci. Technol.* **32**(8), 35-43.

Van Loosdrecht, M. C. M., Picioreanu, C. and Heijnen, J.J. (1997) A more unifying hypothesis for biofilm structures. *FEMS Microbiol. Ecol.* **24**(2), 181-183.

Comparison of the granulation of activated sludge grown on carbohydrate- and protein-rich wastewaters

Han-Qing Yu*, Yu-Ming Zheng, Kui-Zu Su

*Department of Chemistry, The University of Science & Technology of China, Hefei, Anhui, 230026 China (E-mail: hqyu@ustc.edu.cn)

Abstract Granular activated sludge was cultivated in two sequencing batch reactors fed respectively with a synthetic sucrose-rich wastewater and a soybean-processing wastewater. The formation of granular activated sludge in the two reactors and their physicochemical characteristics were compared. With the granulation, the settling ability of the sludge continuously improved, as evidenced by a decreased sludge volume index and an increased settling velocity. The mature granular sludge was nearly spherical and had a strong structure. The granular activated sludge in the two reactors had similar physicochemical characteristics. This suggests that the substrate component was not a crucial factor for the granulation of activated sludge. The data reported in this study provided useful information about the development of aerobic-granule-based bioreactors for the treatment of carbohydrate- and protein-laden wastewaters.

Keywords Activated sludge; Carbohydrates; Granulation; Sequencing batch reactor (SBR); Soybean-processing wastewater; Sucrose

Introduction

After the original activated sludge process was invented, various modified processes have been developed and applied to treatment municipal and industrial wastewaters. Since the performance of the activated sludge processes is limited by the capacity of a clarifier for the separation of the activated sludge from the treated effluent, the self-immobilization of sludge is an attractive option because of its prompting solid–liquid separation in the clarifier. It has been recently reported that the biomass in the sequencing batch reactor (SBR) produces settling granules, which facilitates good solid-liquid separation and the accumulation of high amounts of activated sludge (Morgenroth et al., 1997; Beun et al., 1999; Etterer and Wilderer, 2001).

As compared with conventional activated sludge flocs, the advantages of granular activated sludge are compact and strong in structure (Etterer and Wilderer, 2001). It also has good settle ability and high capacity for biomass retention, and is able to withstand high organic loading rates (Morgenroth et al., 1997). With molasses as the sole carbon source, round-shaped aerobic granules with an average diameter of 0.6 mm started to appear in an SBR after around 40 days of operation. In an aerobic SBR fed with sodium acetate, granules with a diameter of 0.3-0.5 mm were formed after one month of operation (Peng et al., 1999). Large-sized aerobic granules with an average diameter of 3.3 mm were produced in an SBR fed with ethanol and operated at a short hydraulic retention time of 6.75 h and influent chemical oxygen demand (COD) of 830 mg/l. A relative high shear was found favorable for granulation (Beun et al., 1999). Aerobic granules could be formed within 3 weeks in two SBRs fed respectively with glucose and acetate (Tay et al., 2002).

However, so far, information about the physicochemical characteristics of aerobic granules grown on different carbon sources, especially protein, is still sparse. Therefore, the main objective of this work was to explore the granulation process and to compare the physicochemical properties of granules cultivated respectively with sucrose- and protein-rich wastewaters.

Materials and Methods

Experimental Set-Up and Operation

Two SBRs were used in this study. The SBR receiving the synthetic sucrose-rich wastewater was designated as R_C, while another treating the soybean-processing wastewater was designated as R_S. Each reactor had a working volume of 2.2 liters with an internal diameter of 6.0 cm and a height of 90 cm (Fig. 1). Air was introduced through an air diffuser by an air pump at the bottom of the reactor at an air velocity of 2.2 cm/s. The airflow rate was control by a gas-flow controller. The temperature of the reactor was maintained at 25°C.

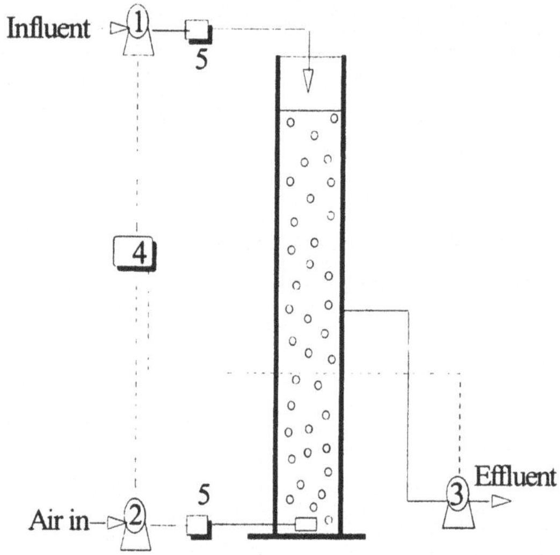

Figure 1: Schematic diagram of the reactor (1.Feed pump; 2.Air pump; 3.Electromagnetism valve; 4.Time controller; 5.Mass-flow controller)

Both reactors were operated in a fill-draw mode. R_C was operated at a loading rate of 3.75 kgCOD/(m^3.d) with a hydraulic retention time of 480 min and influent COD of 1250 mg/l. R_C had successive cycles of 240 min each. One cycle consisted of 10 min of influent addition, 220 min of aeration, 5 min of settling, and 5 min of effluent withdrawal. For R_S, it was operated at a loading rate of 6.00 kgCOD/(m^3.d) with a hydraulic retention time of 240 min and influent COD of 2100 mg/l. It had 5 min of influent filling, 220 min of aeration, 5 min of settling and 10 min of effluent withdrawal. For both reactors, effluent was withdrawn from the port at 50 cm from the reactor bottom with a volumetric ratio of 50%.

Activated sludge from a local municipal wastewater treatment plant was used as inoculum. The inoculum was seeded to each reactor, resulting in initial mixed liquor volatile suspended solids (MLVSS) of 3000 mg/l in R_C and 3400 mg/l in R_S.

Wastewater

The composition of the synthetic wastewater was as follows (in mg/l): sucrose, 830; peptone, 250; beef extract, 160; NH_4Cl, 125; K_2HPO_4, 30; $CaCl_2$, 20; $MgSO_4$, 15; $FeSO_4 \cdot 7H_2O$, 15. The protein-COD accounted for 30% of total-COD. The trace element solution contained (in mg/l): H_3BO_3, 50; $ZnCl_2$, 50; $CuCl_2$, 30; $MnSO_4 \cdot H_2O$, 50; $(NH_4)_6Mo_7O_{24} \cdot 4H_2O$, 50; $AlCl_3$, 50; $CoCl_2 \cdot 6H_2O$, 50 and $NiCl_2$, 50. The influent pH value was adjusted to 7.0 by the addition of $NaHCO_3$ and H_2SO_4.

A protein-laden wastewater, obtained from a local soybean-processing plant, was used to feed R_S. This wastewater had 2100 mg/l of COD, 547 mg/l of proteins, 422 mg/l of carbohydrates and 574 mg/l of total nitrogen. The soluble COD of the wastewater was 1650 mg/l. Since the wastewater contained sufficient amounts of nitrogen, only phosphorus as Na_2HPO_4 was added to ensure the ratio of COD to P to be 100:1. The trace element solution was also dosed.

Analytical Methods

Microbial observation was conducted by using an optical microscope (Olympus CX41). The granular activated sludge size was measured by an image analysis system (Image-pro Express 4.0, Media Cybernetics) with an Olympus CX31 microscope and a digital camera (Olympus C5050 Zoom).

The extracellular polymeric substances (EPS) of sludge were extracted by using EDTA method (Duncan-Hewitt et al., 1989). The carbohydrate concentration in EPS was determined as glucose equivalent using the Dubois method (1956). Protein concentration was measured as bovine albumin equivalent using the Lowry method (1951). The hydrophobic nature of the activated sludge particles was determined by measuring contact angle by axisymmetric drop shape analysis following the method proposed by Duncan-Hewitt et al. (1989).

Measurement of COD, mixed liquor suspended solids (MLSS), MLVSS and sludge volume index (SVI) were performed using the Standard Methods (1992). The settling velocity was measured by recording the time taken for individual granules to fall from a certain height in a measuring cylinder.

Results and Discussion

Granular sludge after 60 days of operation were obtained from the two reactors. Their characteristics were determined in terms of SVI, and mean diameter, settling velocity, hydrophobicity and EPS. These data are summarized in Table 1.

Overall performance of R_C and R_S

As illustrated in Fig. 2A, at the initial operating stage the COD removal efficiency of R_C was about 88%, as the seed sludge had a high bioactivity. With the granulation, the COD removal efficiency increased slightly. It even reached 97% at the end of the experiment. For R_S, its COD removal efficiency was constant around 95%. It kept steadily at 98-99% afterwards.

Table 1: Characteristics of the seed sludge and granular sludge in R_C and R_S

Parameter	Seed sludge	Granules in R_C	Granules in R_S
Diameter (mm)	0.10±0.05	1.01±0.03	1.22±0.09
SVI (ml/g)	74±6	23±4	31±5
Settling velocity (m/h)	7±1	25±4	37±9
Contact angle (°)	40.5±5.3	46.3±4.2	19.3±3.2
Carbohydrate in EPS (mg/g-SS)	3.2±0.2	5.9±0.5	3.1±0.5
Protein in EPS (mg/g-SS)	2.2±0.6	51.4±0.4	4.1±0.8

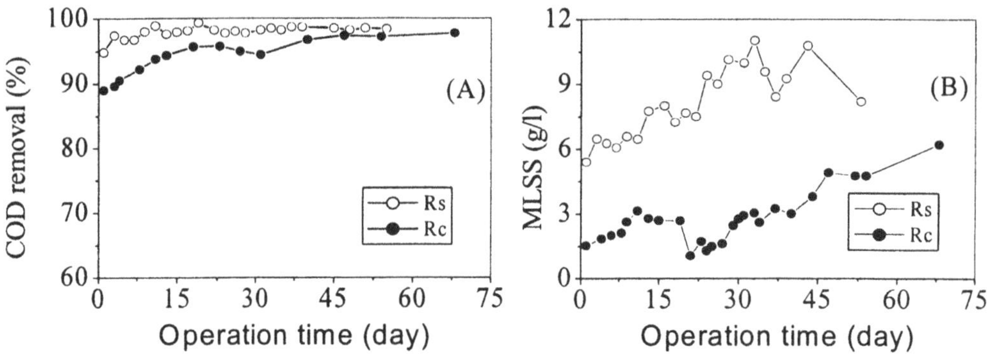

Figure 2: Performance of the two reactors: (A) COD removal efficiency; (B) MLSS

Figure 2B illustrates that after seeding, the biomass concentration in R_C increased slightly. However, on day 20, the MLSS values of the sludge decreased steeply attributed to a disorder of the time controller. After recovery from this failure, the MLSS of the activated sludge slightly rose again, and reached 6.0 g/l as the experiment terminated on day 68. For R_S, a settling period of 10 min was applied in the first week to prevent severe wash out of sludge and then decreased to 5 min in the subsequent 7 days. As a result of the improving settling ability, MLSS kept increasing in spite of the excess sludge wash out. After day 25, the MLSS was stabilized at 9.1 g/l, which was much higher than that of the seed sludge in R_S.

Formation of granules in R_C and R_S

The seed sludge had a fluffy, irregular and loose-structure morphology, as shown in Figs. 3A and 3B. The color of biomass changed from brown to yellow gradually with the process of the experiment.

The activated sludge in R_C showed no significant change in the first week. In the fifth week, tiny granules appeared. From then on, the number of granular sludge increased, and its size increased gradually as well. After 60-day operation, as show in Fig. 3C, granular sludge was matured in R_C. At this stage, R_C was dominated by the mature granular sludge, and little flocs could be observed in this reactor.

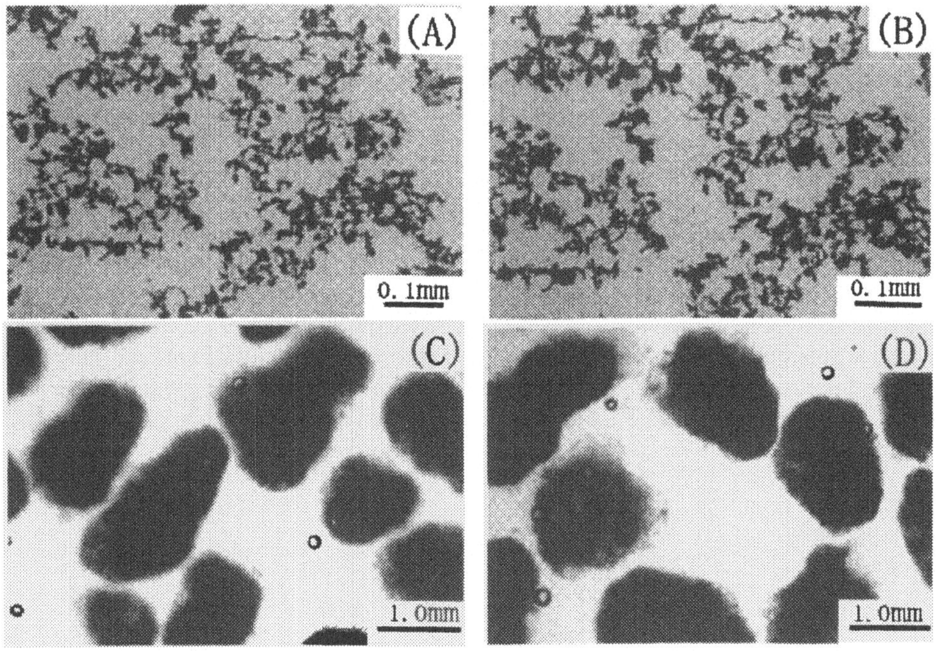

Figure 3: Microscopic observation of (A) seed sludge in R_C; (B) seed sludge in R_S; (C) granular sludge in R_C; (D) granular sludge in R_S

Figure 4: Morphology of (A) granular sludge in R_C; (B) granular sludge in R_S

After two-week operation, in R_S, small granules with diameters of about 0.3-0.5 mm were observed. Thereafter, the number and average diameter of the granules steadily increased. At day 45, granules had a mean diameter of about 1.5 mm.

Figure 4 shows the images of the granular sludge in R_C and R_S on day 60. The granules from the two reactors exhibited a compact structure. As shown in Fig. 4, the granular sludge in R_S had similar morphology to that of the granules in R_C. Figure 5 shows the changes of mean diameter of the granules in the two reactors. The divergence between granule diameters in the two reactors started around day 20. The drop in granule size in R_c coincided with the biomass loss due to time controller disorder. This failure might be partially attributed to the difference between the granule sizes.

In the formation of aerobic granules, it seems to be important to keep a sufficient length of famine period in each cycle. As shown in Fig. 6, within one cycle the COD concentration decreased sharply in the initial 30-min in the two reactors. After that, the COD concentration declined slowly before leveling off at hour 2. The degradation of substrate mainly occurred in the initial 2-h operation. In the latter 2-h operation was a famine period. Such an operating manner was beneficial for the formation of granules (Schwarzenbeck et al., 2004).

Figure 7 illustrates their size distribution. The formation of granules from seed sludge was a gradual process, as evidenced by the increase of mean diameter of the granules. The mean diameter of the granules in the R_S was slightly larger than that of R_C at any given times after their formation.

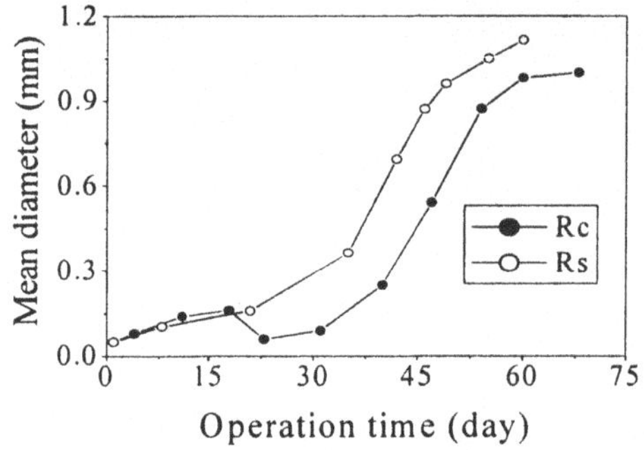

Figure 5: Changes of the sludge mean diameter in the two reactors

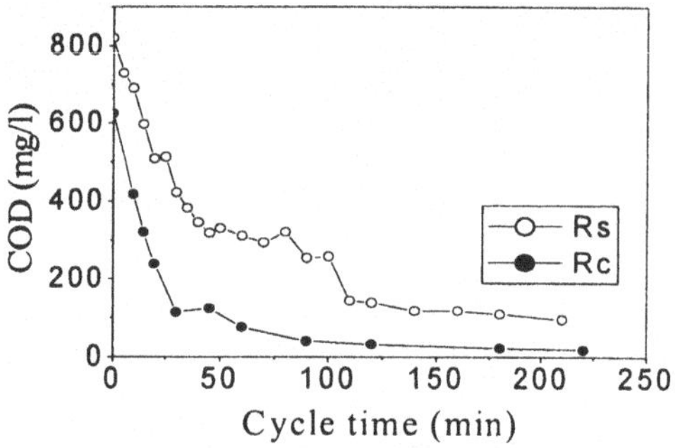

Figure 6: COD profiles within one cycle in the two reactors

Settling ability of the granules

The ratio of MLVSS to MLSS of inoculum was about 66%. After seeding, the ratio increased sharply and reached 0.90 on day 10 in R_C. Thereafter, they were stable between 0.92-0.94. The ratio of MLVSS to MLSS in R_S was 0.87 at the end of 60-day operation. The high MLVSS/MLSS ratios might be attributed to the fact that the inorganic salts concentration in the two wastewaters used in this work was less than municipal wastewater.

The SVI of seed sludge was 74 ml/g. Along with the formation of granules, the SVI decreased gradually. At the termination of the experiment, the SVI of the sludge decreased to only 23 ml/g for R_C and 31 ml/g for R_S, suggesting that the mature granular sludge in the two reactors had an excellent settling capacity.

The average settling velocities of the granular sludge from R_C was in the range of 18-31 m/h, while the corresponding values were 34-42 m/h. This suggests that aerobic granules grown on protein-rich wastewater settled slightly faster than those grown on carbohydrate-rich wastewater.

Hydrophobicity

Hydrophobicity of cell surface plays an important role in the self-immobilization and attachment of cells to a surface (Zita and Hermansson, 1997). As shown in Table 1, a significant difference in sludge hydrophobicity was observed before and after the formation of granules in both R_C and R_S. The contact angle was 35.0° for the seed sludge, but the mean contact angles of the granular sludge from R_C and R_S were 46.3° and 19.5°, respectively. This suggests that the formation of aerobic granular sludge was coupled to a change in the hydrophobicity of the cell surface.

EPS plays a crucial role in maintaining structural integrity in a community of immobilized cells (Liu and Tay, 2001). Table 1 shows that, in R_C, the concentrations of carbohydrate and protein in the EPS of granular sludge were 5.9 and 51.4 mg/gVSS, respectively, and that the corresponding values in the seed sludge were 3.2 and 2.2 mg/gVSS, respectively. The increase in protein concentration was significant. In R_S, the concentrations of carbohydrate and protein in the EPS of granular sludge were 3.1 and 4.4 mg/gVSS, respectively, not substantially different from those in the seed sludge. In both cases, there was no substantial change of carbohydrate concentration between the seed and granular sludges. However, with granulation, the sludge R_C in increased significantly. On the other hand, the protein concentration in EPS of two types of granules was considerably different. This hinters that protein might play a more important role in the formation of granules than carbohydrate.

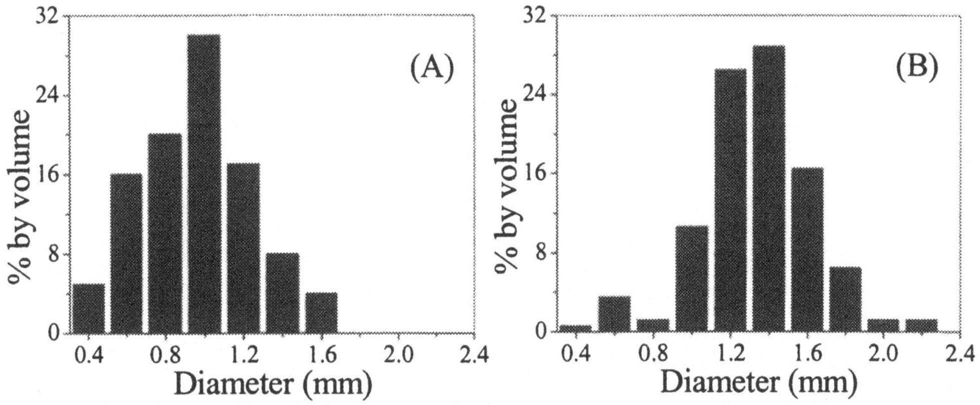

Figure 7: Size distribution of the granules in (A) R_C; (B) R_S

Evaluation of the granules in R_c and R_s

In all previous studies concerning aerobic granulation, various synthetic wastewaters rich in carbohydrates (Etterer and Wilderer, 2001), or fatty acids (Tay et al., 2002), have been used as a carbon source. So far, information about aerobic granulation with an actual industrial wastewater is very limited (Schwarzenbeck et al., 2004). In the present work, aerobic granules were cultivated in the SBR treating soybean-processing wastewater at 25±1°C. This wastewater was rich in proteins with a ratio of proteins:COD of 0.26:1. The aerobic granules grown on this wastewater had excellent settling ability and bioactivity. Their characteristics were similar to those grown on carbohydrate-rich wastewater. This hints that the substrate component was not a crucial factor for the granulation of activated sludge. The information provided here would be useful for the development of aerobic-granule-based bioreactors for the treatment of industrial wastewaters.

Conclusions

This study demonstrated that granular activated sludge could be cultivated in two sequencing batch reactors respectively fed with a synthetic sucrose-rich wastewater and a soybean-processing wastewater. After 60-day operation, stable granules with an average size larger than 1.0 mm were formed. With the granulation, the settling ability of the sludge continuously improved, as evidenced by a decreased sludge volume index and an increased settling velocity. The mature granular sludge was nearly spherical. The granular activated sludge in the two reactors had similar physicochemical characteristics, and had an excellent settling ability. This suggests that the substrate component was not a key factor in the granulation of activated sludge.

Acknowledgements

Authors wish to thank the Trans-Century Training Program Foundation for the Talents, Ministry of Education, China, and the Anhui Foundation for Excellent Talents, China, for the partial financial support of this study.

References

APHA (1992) Standard Methods for the Examination of Water and Wastewater. 18th ed. APHA/AWWA/WEF, Washington, D. C.

Beun J. J., Hendriks, A., van Loosdrecht M. C. M., Morgenroth E., Wilderer P. A. and Heijnen J. J. (1999). Aerobic granulation in a sequencing batch reactor. Wat. Res., 33, 2283-2290.

Dubois M., Gilles K. A., Hamilton J. K., Rebers P. A. and Smith F. (1956). Calorimetric method for determination of sugars and related substances. Anal. Chem., 28, 350–356.

Duncan-Hewitt W. C., Policova Z., Cheng P., Vargha-Butler E. I. and Neumann A. W. (1989). Semi-automatic measurement of contact angles on cell layers by a modified axis symmetric drop shape analysis. Colloids Surfaces, 42, 391–402.

Etterer T. and Wilderer P.A. (2001). Generation and properties of aerobic granular sludge. Wat. Sci. Tech., 43, 19-26.

Liu Y. and Tay J. H. (2001). Detachment forces and their influence on the structure and metabolic behavior of biofilms. World J. Microb. Biotech. 17, 111-117.

Lowry O. H., Rosebrough N. J., Farr A. L. and Randall R. J. (1951). Protein measurement with the Folin phenol reagent. J. Biol. Chem., 193, 265-275.

Morgenroth E., Sherden T., van Loosdrecht M. C. M., Heijnen J. J. and Wilderer P. A. (1997) Aerobic granular sludge in a sequencing batch reactor. Wat. Res., 31, 3191-3194.

Peng D. C., Bernet N., Delgenes J. P. and Moletta R. (1999) Aerobic granular sludge - A case report. Wat. Res., 33, 890-893.

Schwarzenbeck N., Erley R., Swain B. S. M., Wilderer P. A. and Irvine R. L. (2004) Treatment of malting wastewater in a granular sequencing batch reactor. Acta. Hydrochim. Hydrobiol., 32, 16-24.

Tay J. H., Liu Q. S. and Liu Y. (2002) Characteristics of aerobic granules grown on glucose and acetate in sequential aerobic sludge blanket reactors. Environ. Tech., 23, 931-936.

Zita A. and Hermansson M. (1997) Effects of bacterial cell surface structures and hydrophobicity on attachment to activated sludge flocs. Appl. Environ. Microbiol., 63,1168-1170.

Characteristics and Applicability of Nitrifying Granules Produced in an Aerobic Upflow Fluidized Bed Reactor

S. Tsuneda*, Y. Ejiri*, M. Ogiwara*, T. Nagano* and A. Hirata*

*Department of Chemical Engineering, Waseda University, 3-4-1 Ohkubo, Shinjuku-ku, Tokyo 169-8555, Japan

Abstract A new technique for producing nitrifying granules using an aerobic upflow fluidized bed (AUFB) reactor was proposed. It was found that pre-aggregation of seed sludge using hematite fine particles promoted production of nitrifying granules. Fluorescence *in situ* hybridization (FISH) analysis visualized the change in the spatial distribution of nitrifying bacteria in the granules, demonstrating that small aggregates of nitrifying bacteria play roles of cores for further growth to nitrifying granules. To show the high nitrification potential of the granules, pure oxygen was supplied to the AUFB reactor. The resulting ammonia removal rate reached 16.7 kg-N/m^3/day with a sustained ammonia removal efficiency of over 80%. In another experiment, when the granules were used in a continuous stirring tank reactor (CSTR), a high biomass density of nitrifying bacteria was retained in the reactor even under short HRT (24 min) conditions. This high performance is attributed to the special characteristics that the settling velocity of the nitrifying granules of a diameter of 1500 μm reach as high as 1.4 - 1.5 cm/s, which is about 50 times higher than the liquid velocity at the exit of the CSTR at this HRT.

Keywords Aerobic upflow fluidized bed; hematite; nitrification potential; nitrifying granule; pure oxygen; short HRT

Introduction

Since autotrophic nitrifying bacteria hardly form biofilms on carriers due to their extremely low growth rate and lack of production of extracellular polymeric substances, it is generally accepted that retaining a large amount of nitrifying bacteria in a reactor is difficult to achieve (Tsuneda *et al.*, 2001). Granulation is thought to be an effective technique for immobilizing nitrifying bacteria. Although there have been various reports regarding aerobic granulation in organic wastewater treatment processes (Mishima and Nakamura, 1991; Beun *et al.*, 1999; Tay *et al.*, 2001), granulation of nitrifying bacteria in inorganic wastewater treatment processes has only been observed in a small number of reactor types, such as front-aeration (de Beer *et al.*, 1993) or sequencing batch reactors (Tay *et al.*, 2002).

This paper introduces a new technique for producing nitrifying granules using an aerobic upflow fluidized bed (AUFB) reactor (Tsuneda *et al.*, 2003) where inorganic wastewater containing 500 g/m^3 of ammonia nitrogen is continuously fed. The influence of sludge pre-aggregation on nitrifying granulation rate is investigated and a mechanism for the formation of nitrifying granules in the AUFB reactor is discussed. The sedimentation characteristics and microbial distribution of the granules with different sizes are then examined. In this work, in order to show the nitrification potential of these granules, a high-ammonia-loading continuous nitrification experiment was conducted by supplying a high concentration of oxygen gas to the AUFB reactor. Furthermore, in order to clarify the utility of the nitrifying granules produced in the AUFB reactor, they were transferred to a continuous stirring tank reactor (CSTR) and applied to the nitrification of inorganic wastewaters with low ammonia concentration (30 - 50 g-N/m^3) under short hydraulic retention time (24 - 72 min).

Materials and methods

AUFB reactor structure and operating conditions

The scheme of the AUFB reactor used in this study is shown in Figure 1. The AUFB reactor is a column with a diameter of 50 mm, a height of 3.2 m and an effective volume of 6.3 L. From the bottom of the reactor, synthetic inorganic wastewater was continuously fed. A solid-liquid separator was placed on the top of the reactor to prevent washout of sludge and granules from the reactor. The synthetic wastewater was composed of 2360 g/m^3 of $(NH_4)_2SO_4$ (500 g/m^3 as NH_4^+-N basis) and 6040 g/m^3 of Na_2SO_4 with the addition of 1 g-P/m^3 of KH_2PO_4 and 1 g-Fe/m^3 of $FeSO_4 \cdot 7H_2O$ as trace materials and 1000 g-C/m^3 of $NaHCO_3$ as pH control and an inorganic carbon source for the nitrifying bacteria. Aeration was carried out via a porous air diffuser ball at a rate of 0.16 L/min/L-bed. The initial ammonia loading rate was 0.5 kg-N/m^3/day, and then it was raised by the reducing hydraulic retention time (HRT) gradually while confirming a stable ammonia removal efficiency.

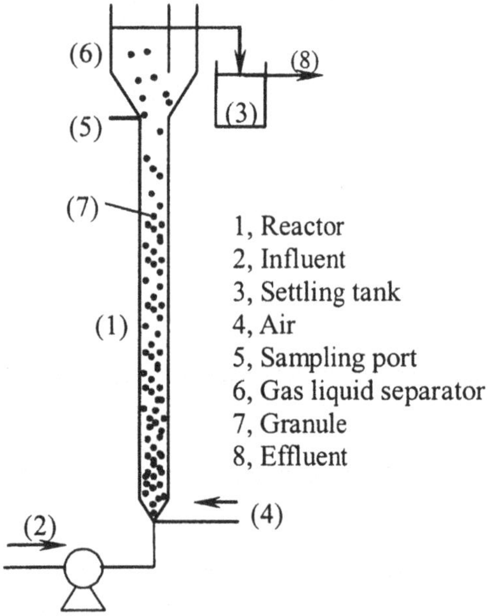

Figure 1: Schematic illustration of the AUFB reactor used in the experiment.

Seed sludge was obtained from an aerobic basin of a municipal wastewater treatment plant and was acclimated to the above-described synthetic inorganic wastewater for two years. In the first run (Run 1), this seed sludge was inoculated to the AUFB reactor without any pre-treatment. In the second run (Run 2), seed sludge was pre-aggregated with hematite (Fe_2O_3) fine particles before being inoculated to the AUFB reactor. Hematite particles were prepared by aging a solution containing 180 mmol/L $FeCl_3$ and 1.0 mmol/L HCl for 24 h at 100°C, as described by Matijevic and Scheiner (1978). The mean diameter of individual particle was determined to be 105 nm by transmission electron microscope. The pre-aggregation procedure was as follows: hematite, the dry weight of which was 1/30 of seed sludge, was added to the seed sludge and mixed together until a significant portion of the sludge was trapped by hematite, which was confirmed by observation with a light microscope (VH-Z450, Keyence, Japan), following which the sludge was incubated in the inorganic wastewater using a shaker for a day before being inoculated to the AUFB reactor.

Fluorescence *in situ* hybridization (FISH)

The fixation and sectioning procedures of a granule sample to be applied to FISH followed those of Aoi *et al.* (2000). The following 16S rRNA-targeted oligonucleotide probes were used: NEU23a labelled with FITC, specific for a region in the 16S rRNA of ammonia-oxidizing bacteria which are halophilic and halotolerant members of the genus *Nitrosomonas* (Wagner *et al.*, 1995); and EUB338 labelled with Cy3, general for all eubacteria (Amann *et al.*, 1990). Hybridization and washing were performed according to the method of Aoi *et al.* (2000). The granule sections hybridized with the probes were observed under a fluorescence microscope (Axioskop 2, Zeiss, Germany).

Settling velocity

Nitrifying granules with different diameters (500 - 1900 μm) were prepared. An individual granule was carefully placed in a cylinder filled with water to a height of 50 cm, and then the settling velocity of each granule was estimated based on the time taken to reach the bottom.

Nitrification under supply of high concentration of oxygen

Nitrifying granules with a diameter of 1500 μm were added to an AUFB reactor with a diameter of 31 mm, a height of 45 cm and an effective volume of 0.34 L. The initial MLSS concentration was 15000 g/m^3. From the bottom of the reactor, the same wastewater (NH_4^+-N: 500 g/m^3) as used in the granulation process was continuously fed. Air (0 - 35.5 h), 40% O_2 gas (35.5 - 97.0 h), 60% O_2 gas (97.0 -182 h), 80% O_2 gas (182 - 255 h) and 100% O_2 gas (255 - 336 h) were successively supplied to the reactor via a glass filter (11GP16, Sibata, Japan). The HRT was initially 3.2 h. When the effluent concentration reached a steady state, the HRT was reduced in a stepwise manner to elucidate the maximum nitrification rate under supply of each concentration of oxygen.

Nitrification under short HRT conditions

Nitrifying granules with a diameter of 1500 μm were added to a continuous stirring tank reactor (CSTR) whose effective volume was 0.57 L. The initial MLSS concentration was 12700 g/m^3. The synthetic wastewater (NH_4^+-N: 500 g/m^3) was diluted ten times and used as the initial influent solution (NH_4^+-N: 50 g/m^3) for the nitrification test under short HRT conditions. The $NaHCO_3$ used as the inorganic carbon source maintained pH within the range of 7.0 - 7.3 throughout the experiment. The volumetric aeration rate was set to 1.0 L/min and the dissolved oxygen (DO) concentration kept between 4 and 6 g/m^3. The HRT was initially 72 min, and was reduced in a stepwise manner during the experiment when the effluent concentration reached a steady state. When the nitrification efficiency reached the limit, the dilution ratio of wastewater was raised with reducing HRT so as to maintain a constant ammonia loading rate (1.86 kg-N/m^3/day).

Chemical analysis of water quality

All samples obtained from the reactor were filtered with a glass filter (GF/F, Whatman, UK), and then the concentration of NH_4^+-N was measured by ion chromatography (DX-120, Dionex, Japan). Measurement of MLSS followed the standard method (APHA, 1995).

Evaluation of granule size

The average diameter of granules was determined using an optical microscope (VH-Z450, Keyence, Japan) to measure the diameter of 100 granules randomly obtained from the reactor. A circler shape was assumed for the granules in obtaining the diameter.

Results and discussion

Formation of nitrifying granules in the AUFB reactor

Young granules were observed at the bottom of the AUFB reactor around day 60 for both Runs 1 and 2. These granules increased in size and became spherical, pseudocubic or elliptical in shape. SEM images of the granules sampled from the AUFB reactor (Run 1) on day 200 are shown in Figure 2. It was confirmed that the bacteria formed dense layers on the granule surface.

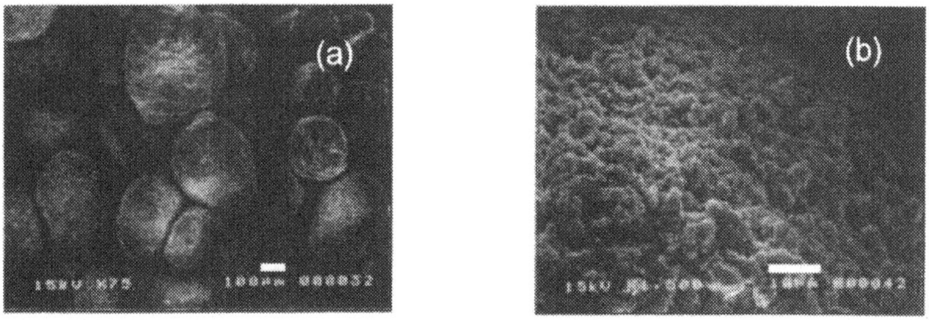

Figure 2: SEM images of the nitrifying granules in the AUFB reactor. (a) external appearance of granules, bar length: 100 µm; (b) surface morphology of granules, bar length: 10 µm.

Two granulation experiments were conducted: Run 1, without pre-aggregation of seed sludge and Run 2, with pre-aggregation of seed sludge. The transitions of the average granule diameter in both runs are shown in Figure 3. These results confirm the earlier growth of nitrifying granules in Run 2, in which pre-aggregated sludge was inoculated. Since the only difference between Runs 1 and 2 was the occurrence of pre-aggregation of seed sludge, and other operational conditions were the same, it became evident that pre-aggregation of seed sludge promotes nitrifying granulation. On the other hand, Figure 3 shows that pre-aggregation by hematite is not a necessary action towards nitrifying granulation. This implies that 1 g-Fe/m^3 of $FeSO_4 \cdot 7H_2O$ as trace materials is enough as inorganic nucleus for nitrifying granulation. Since the size of hematite particles used in this study is smaller than bacterial cells, hematite plays a role of chemical coagulant to assist floc formation in a similar manner as talc (Clauss *et al.*, 1999). This assumption was supported by the result that the MLVSS/MLSS values of the granules produced in Run 2 were lower than those produced in Run 1 in early stages, and then, both runs exhibited the same MLVSS/MLSS value after 150 days (data not shown). This result indicated that younger granules in Run 2 contained hematite in their cores, which might promote further granulation processes. However, after 150 days, if two granules with the identical size were picked up from Runs 1 and 2, no difference in structure, settling ability and nitrification performance was observed.

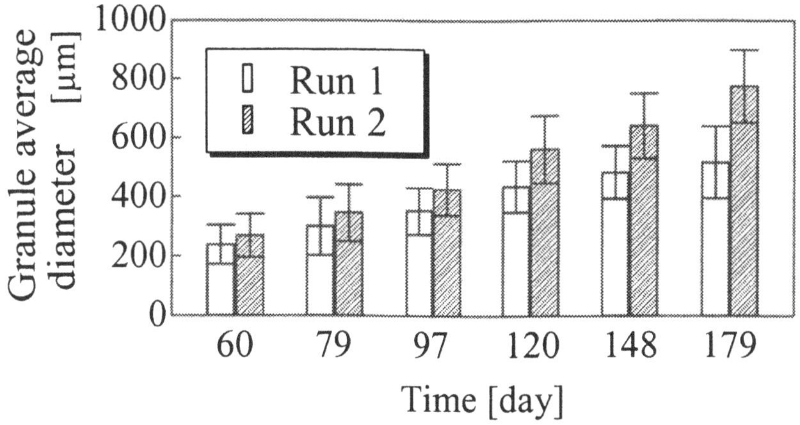

Figure 3: Effect of pre-aggregation of seeding sludge on the growth rate of granules. Run 1: without pre-aggregation; Run 2: with pre-aggregation.

Microbial distribution in nitrifying granules

The FISH images of the nitrifying granules obtained using NSO190 a genus-specific probe for the 16S rRNA region of all ammonia-oxidizing bacteria (AOB) belonging to beta subclass of *Proteobacteria* were completely identical with those obtained using NEU23a (data not shown), which revealed that the genus *Nitrosomonas* (together with *Nitrosococcus mobilis*) were the prominent AOB in the granules. The time course of FISH images of the granules stained by NEU23a and EUB338 probes is shown in Figure 4. On days 65 and 92, many clusters of AOB were observed throughout the granules, indicating cell distribution in all parts of the young granule (< 500 μm). However, on day 178, cell distribution was limited to within 100 μm of the surface in the mature granule (> 600 μm). Other bacteria detected only by EUB338 probe were significantly fewer than AOB, probably because the lack of organic compounds limited the growth of these bacteria. Figure 4 also gives information with regard to changes in the appearance of nitrifying granules. Porous, irregular shaped granule-like aggregated sludge was observed on day 65. Over time, the surfaces of granules became smooth, and their shapes were stabilized in dense spherical or elliptical forms, as shown by Figure 4c. This is probably due to increased collision and frictional forces between granules in the AUFB reactor with increasing granule diameter. The dense structure probably limits oxygen diffusion, resulting in inactive zone inside the granules as observed by FISH analysis (Figure 4c).

From these results, the following formation process of nitrifying granules is proposed: 1) first, nitrifying bacteria aggregate to produce granule cores; 2) second, the aggregates grow to spherical or elliptical forms due to multiplication of the nitrifying bacteria under moderate shear stress in the reactor, resulting in mature nitrifying granules.

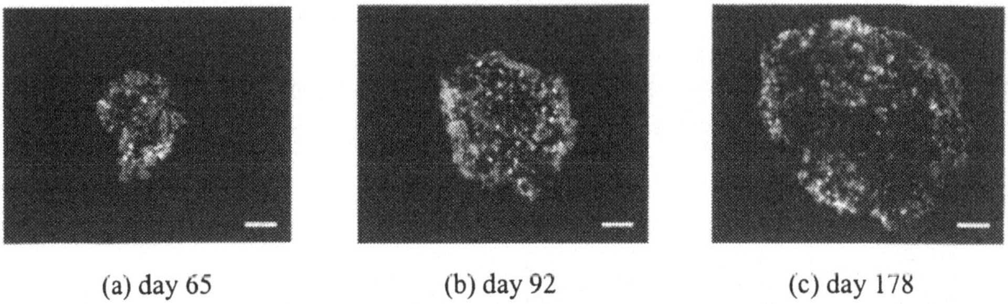

Figure 4: *In situ* hybridization of granule section with pre-aggregation viewed by fluorescence microscopy; white part: ammonia-oxidizing bacteria (NEU23a and EUB338); grey part: other bacteria (EUB338); bar length: 100 µm. Light grey colour found in the centre of the large granules is derived from self-fluorescence or non-specific hybridization, and thus this zone is biologically inactive.

Settling velocity of nitrifying granules

Figure 5 shows the settling velocities of nitrifying granules with different sizes. Up to a granule diameter of less than 1500 µm, the settling velocity increased linearly with increasing diameter. However, above this diameter, the settling velocity remained almost constant at 1.4 - 1.5 cm/s.

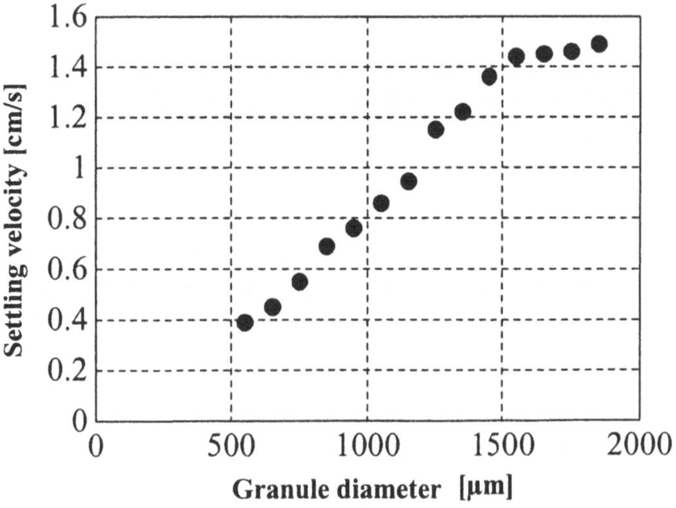

Figur 5: Settling velocity of nitrifying granules with various diameters.

Taking into account the FISH result that the cell distribution was limited to within 100 μm of the surface in the mature granule (> 600 μm) owing to limitation of oxygen diffusion, the nitrifying granules of diameter more than 1500 μm provides much inactive zone in spite of almost identical settling velocity, and thus those granules might be unfavourable for practical wastewater treatment. Crashing such large nitrifying granules into small pieces using a homogenizer might be one of the solutions to reduce inactive zone inside the granules.

Nitrification efficiency under supply of high concentration of oxygen

In this experiment, in order to show the potential of nitrifying granules, a high concentration of ammonia (500 g-N/m^3) in the liquid and a high concentration of oxygen (20 - 100%) in the gas were supplied to the AUFB reactor. Time courses of the ammonia removal rate and removal efficiency are shown in Figure 6, where I - V indicate oxygen partial pressures in the aeration gas of 20% (air), 40%, 60%, 80% and 100% (pure oxygen), respectively. When the effluent concentration reached a steady state, the HRT was reduced in a stepwise manner to elucidate the maximum nitrification rate under supply of each concentration of oxygen. When supplying pure oxygen, ammonia removal rate reached 16.7 kg-N/m^3/day while sustaining an ammonia removal efficiency of more than 80%. This result indicates that retaining a high biomass density of nitrifying bacteria in the reactor in the form of granules enables high-rate nitrification providing that sufficient oxygen is supplied.

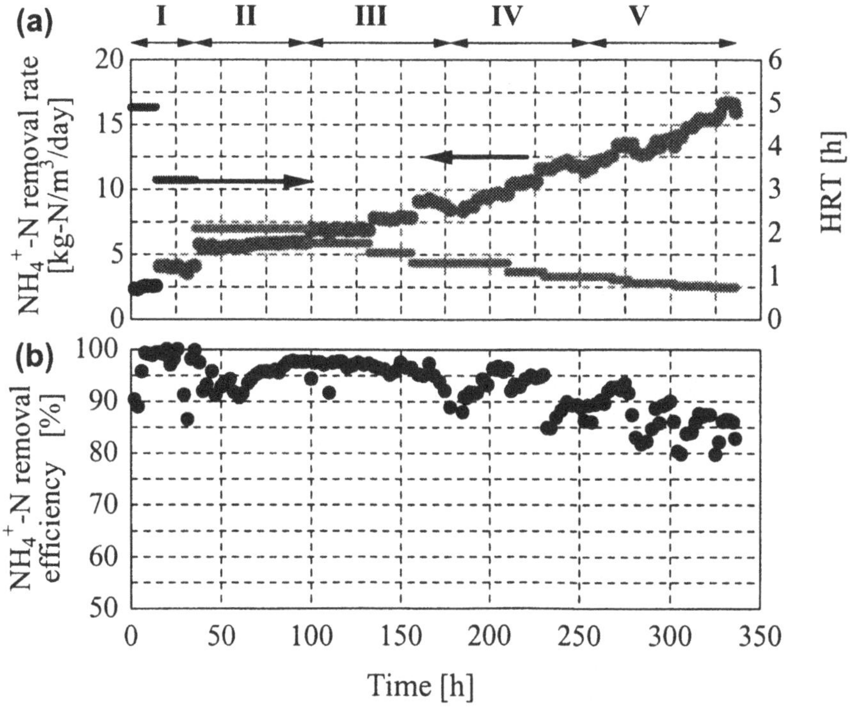

Figure 6: Time course of (a) ammonia removal rate and (b) ammonia removal efficiency over step changes in concentration of oxygen. The partial pressures of oxygen in the supplied gas were as follows. I: Air (0 - 35.5 h), II: 40% O_2 gas (35.5 - 97.0 h), III: 60% O_2 gas (97.0 -182 h), IV: 80% O_2 gas (182 - 255 h) and V: 100% O_2 gas (255 - 336 h).

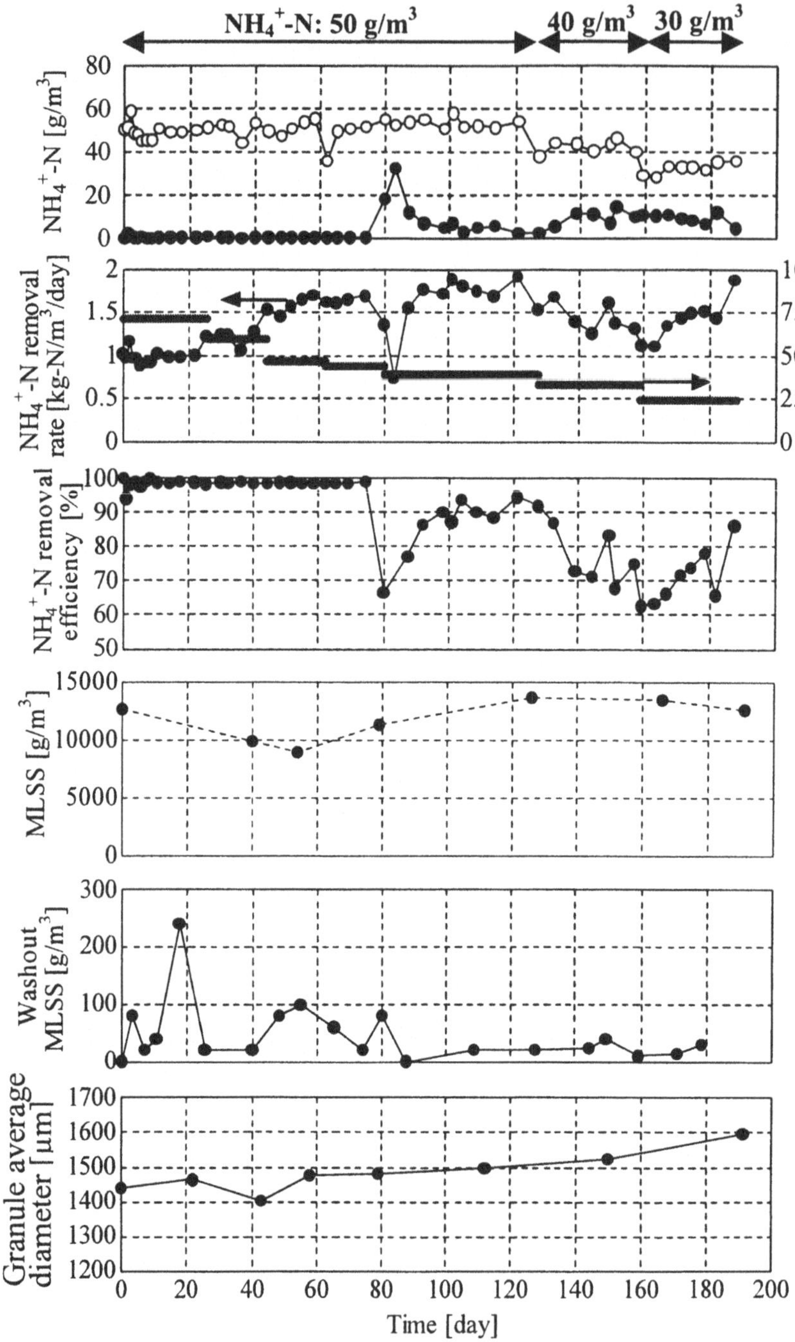

Figure 7: Time course of effluent ammonia concentration, ammonia removal rate and efficiencies, MLSS in the CSTR, washout MLSS and granule average diameter undergoing nitrification under short HRT conditions. Open circles in the top graph indicate NH_4^+-N concentration in the influent.

Nitrification efficiency under short HRT conditions

In order to show further potential of the nitrifying granules, they were applied to wastewater with a low concentration of ammonia (50 g-N/m^3) under short HRT conditions (24 - 72 min). The time courses of water quality data (influent and effluent concentrations of NH_4^+-N, granule average diameter, MLSS and washout MLSS) are shown in Figure 7 along with the ammonia removal rate and efficiency. Complete nitrification was attained and kept stable even when the HRT was as short as 44 min. The ammonia removal rate reached 1.65 kg-N/m^3/day on day 75. However, when the HRT was further shortened to 38 min, the ammonia removal efficiency suddenly decreased. When the efficiency recovered to 90% (on day 126), influent ammonia concentration was decreased to 40 g/m^3 and the HRT was simultaneously shortened to maintain an ammonia loading rate of 1.86 kg-N/m^3/day. However, the ammonia removal efficiency never recovered. Further reduction of ammonia concentration to 30 g/m^3 failed to recover the ammonia removal efficiency. DO concentration in the reactor was over 4.0 g/m^3 throughout the experiment, indicating that ammonia removal efficiency was not limited by substrate but limited by each cell potential in the granule.

Washout of biomass was hardly observed except during the initial 50 days, indicating that nitrifying bacteria could be completely retained in the reactor even if the HRT was as short as 24 min. This result can be explained by comparing the liquid velocity with the settling velocity of the granules. When the HRT is 24 min, the liquid velocity at the exit of the reactor is 0.027 cm/s. As shown in Figure 5, granules of a diameter of 1500 μm exhibited a settling velocity of 1.4 cm/s, which is about 50 times higher than the liquid velocity at this HRT. Therefore, it was demonstrated that the nitrifying granules can be applied to high-rate nitrification under short HRT conditions.

Conclusions

In this study, the conditions of nitrifying granulation in the AUFB reactor were elucidated, and then the characteristics and potential of the nitrifying granules were examined. The following results were obtained.

(1) Nitrifying granulation was observed in the AUFB reactor when inorganic wastewater containing 500 g/m^3 of ammonia nitrogen was continuously fed with a volumetric aeration rate of 0.16 L/min/L-bed.

(2) Pre-aggregation of seed sludge using hematite fine particles promoted formation of nitrifying granules.

(3) FISH analysis revealed that ammonia-oxidizing bacteria existed uniformly throughout the younger granule (< 500 μm), whereas their presence was limited to within 100 μm of the surface in the mature granule (> 600 μm) probably because of limitation of oxygen diffusion into the centre of the dense granules.

(4) The settling velocity of nitrifying granules increased with increasing granule size (< 1500 μm), but was constant (1.4 - 1.5 cm/s) for larger granules (> 1500 μm).

(5) When pure oxygen was supplied to the AUFB reactor, the ammonia removal rate reached 16.7 kg-N/m^3/day with a sustained ammonia removal efficiency of more than 80%.

(6) Nitrifying granules can be retained in a CSTR at a high density even under short HRT (24 min) conditions, because their settling velocity is about 50 times higher than the liquid velocity at the exit of the CSTR.

These results help to reveal the mechanism of nitrifying granulation in the AUFB reactor and demonstrate the applicability of nitrifying granules to high-throughput nitrification processes in various types of reactors under various operational conditions.

References

APHA (1995). *Standard Methods for the Examination of Water and Wastewater.* 19th edn, American Public Health Association/American Water Works Association/Water Environment Federation, Washington DC, USA.

Amann, R.I., Krumholz, L. and Stahl, D.A. (1990) Fluorescent-oligonucleotide probing of whole cells for determinative, phylogenetic, and environmental-studies in microbiology. *J. Bacteriol.* **172**(2), 762–770.

Aoi, Y., Miyoshi, T., Okamoto, T., Tsuneda, S., Hirata, A., Kitayama, A. and Nagamune, T. (2000) Microbial ecology of nitrifying bacteria in wastewater treatment process examined by fluorescence in situ hybridization. *J. Biosci. Bioeng.* **90**(3), 234–240.

Beun, J.J., Hendriks, A., van Loosdrecht, M.C.M., Morgenroth, E., Wilderer, P.A. and Heijnen, J.J. (1999) Aerobic granulation in a sequencing batch reactor. *Water Res.* **33**(10), 2283–2290.

Clauss, F., Balavoine, C., Helaine, D. and Martin, G. (1999) Controlling the settling of activated sludge in pulp and paper wastewater treatment plants. *Water Sci. Technol.* **40**(11-12), 223–229.

de Beer, D., van den Heuvel, J.C. and Ottengraf, S.P.P. (1993) Microelectrode measurement of the activity distribution in nitrifying aggregates. *Appl. Environ. Microbiol.* **59**(2), 573–579.

Matijevic, E. and Scheiner, P. (1978) Ferric hydrous oxide sols III. Preparation of uniform particles by hydrolysis of Fe(III)-chloride, -nitrate, and –percholates solutions. *J. Colloid Interface Sci.*, **63**(3), 509–524.

Mishima, K. and Nakamura, M. (1991) Self-immobilization of aerobic activated-sludge –A pilot-study of the aerobic upflow sludge blanket process in municipal sewage-treatment. *Water Sci. Technol.* **23**(4-6), 981–990.

Tay, J.H., Liu, Q.S. and Liu, Y. (2001) Microscopic observation of aerobic granulation in sequential aerobic sludge reactor. *J. Appl. Microbiol.* **91**(1), 168–175.

Tay, J.H., Yang, S.F. and Liu, Y. (2002) Hydraulic selection pressure-induced nitrifying granulation in sequencing batch reactors. *Appl. Microbiol. Biotechnol.* **59**(2-3), 332–337.

Tsuneda, S., Park, S., Hayashi, H., Jung, J. and Hirata, A. (2001) Enhancement of nitrifying biofilm using selected EPS produced by heterotrophic bacteria. *Water Sci. Technol.* **43**(6), 197–204.

Tsuneda, S., Nagano, T., Hoshino, T., Ejiri, Y., Noda, N. and Hirata, A. (2003) Characterization of nitrifying granules produced in an aerobic upflow fluidized bed reactor. *Water Res.* **37**(20), 4965–4973.

Wagner, M., Rath, G., Amann, R.I., Koops, H.P. and Schleifer, K.H. (1995) In situ identification of ammonia-oxidizing bacteria. *Syst. Appl. Microbiol.* **18**(2), 251–264.

Aerobic granulation during the start up period of a periodic biofilter

C. Di Iaconi* , R. Ramadori, A. Lopez***

* Istituto di Ricerca Sulle Acque C.N.R - Via F. De Blasio 5 – 70123 Bari (Italy)
** Istituto di Ricerca Sulle Acque C.N.R. - Via Reno 1 – 00198 Roma (Italy)

Abstract The paper reports the results of an experimental investigation specifically aimed at assessing the mechanism of biomass granulation during the start up of a period biofilter under aerobic conditions. The role of the hydrodynamic shear forces, calculated by a methodology developed by authors, was assessed, and a mechanism for granulation was proposed: (1) formation of a thin biofilm that completely covers the carrier; (2) increase of biofilm thickness; (3) break-up of the attached biofilm with release of biofilm particles; (4) rearrangement of biofilm particles in smooth granules. The hydrodynamic shear forces trend during the start-up period provides a key for explaining the process of biomass granulation. In fact, during the first two steps, the biofilter is characterised by rather weak shear force values (lower than 1 dyne/cm2). Under these weak forces, the biofilm grows, increasing its thickness through a porous structure characterized by weak adhesion strengths. The continual increase of biofilm thickness produces a corresponding increase of the shear forces with negative effects on biomass stability that cause the detachment of biofilm particles. In turn, such detachment causes a further sharp increase of the shear forces that promotes the rearrangement of the detached biofilm particles into smooth granules.

Keywords. Aerobic granulation, hydrodynamic shear force, SBBR, start up.

Introduction

The goals of any biological wastewater treatment are essentially two: high volumetric conversion capacity and low biomass production rates. The conversion capacity is directly proportional to the biomass growth rate via the yield factor (Y), whereas the biomass production is determined by the difference between the actual growth rate (μ) and the biomass losses due to endogenous metabolism, death, predation and lysis.

In aerobic processes, considering that the yield factor is rather high (Y= 0.5-0.7), the systems designed at growth rates close to the maximum value (μ_{max}=0.3-0.6 h^{-1}), such as biofilters, are characterized by high volumetric conversion capacity (6-8 kg COD/m^3·d) as well as high biomass production rates (0.4-0.5 kg TSS/kg $COD_{removed}$, Pujol *et al.*, 1992). To get low biomass production, the biomass growth rate must be reduced close to maintenance requirements. However, in this manner the volumetric conversion capacities will be reduced unless the biomass retention is increased. In activated sludge systems, because of slow biomass settling velocity, biomass retention cannot be increased beyond a fixed value. For such a reason, these systems are characterised by low volumetric conversion capacity (0.5-1.0 kg COD/m^3·d) and need large reaction volumes to guarantee sufficient treatment capacity. Consequently, as aeration basins cannot be very deep, large reaction volumes are obtained increasing the superficial area with negative effects in terms of odours, noise and aerosols.

In any case, although presently conventional suspended biomass reactors and biofilters are the two most frequently used systems, associated environmental impact problems and the high cost of land in populated areas will force their progressive replacement and/or upgrading with innovative processes characterized by high performances, plant compactness, reduced sludge production and lower costs. Among the new technologies recently proposed to satisfy such requirements, the most promising are aerobic systems with granular biomass. In such systems, relatively high biomass concentrations (up to 15 g/l) and conversion capacities (6-7 $kg_{COD}/m^3 \cdot d$) are achieved allowing rather low sludge production rates (Mulder *et al.*, 2001).

Previous experiences show that operational periodic conditions are crucial for the granular growth of biomass. In fact, until now, aerobic granulation has been observed only in Sequencing Batch Reactors (SBRs) or, more recently, in Sequencing Batch Biofilm Reactors (SBBRs), i.e. submerged filters that operate in a "fill and draw" mode (Di Iaconi *et al.*, 2004). Presently, two main strategies are followed to achieve aerobic granulation:

- spontaneous aerobic granulation of suspended growth has been obtained in the traditional SBR applying short fill periods and short settling times (Morgenroth *et al.*, 1997; Etter and Wilderer, 2001; McSwain *et al.*, 2004).
- aerobic granulation has been stimulated by appropriate hydrodynamic shear forces (Beun *et al.*, 2002; Liu and Tay., 2002).

Following this second approach it has been pointed out that a certain value of hydrodynamic shear forces is required for the formation of aerobic granules although the mechanism by which these forces influence the formation, structure and metabolism of granular sludge is not fully understood yet. In any case, it is interesting to observe that though hydrodynamic shear forces are considered as one of the most decisive factors in the formation and stability of granular biomass, to the authors' knowledge no papers specifically dealing with the calculation of such forces have been published to date.

In this framework, the present paper reports the results of an experimental study specifically aimed at developing a simple methodology for calculating hydrodynamic shear forces in SBBRs. Then, by applying such a methodology during the start-up period, plausible mechanism for biomass granulation is accessed.

Materials and methods

Experimental reactor and operation

Granular sludge was grown in a laboratory scale SBBR shown in figure 1 whose features are reported in table 1. The reactor was a plexiglass closed cylindrical vessel filled with biomass supporting material (KMT-k1 elements from Kaldnes - Norway) kept between two sieves and aerated by air injection through porous stones placed close to the upper sieve. Here, the dissolved oxygen concentration was held above 4 mg/L. The reactor was equipped with an external loop for recirculating wastewater.

The start up period lasted approximately 3 months during which the organic load was progressively increased from 0.5 to 4 $kg_{COD}/m^3 \cdot d$. The SBBR was operated with cycles of 8 h (6 min for influent addition, 464 min for aeration, 10 min for effluent withdrawal). During the start-up period, no washing operations were carried out, the headlosses at the bottom of the reactor increased from 0.1 m to 2 m, and the biomass attached to the carrier changed its morphology gradually but continuously. After this period, the headlosses in the reactor reached values so high that a first washing step was required. This operation was carried out using compressed air (1.5 bar) for a few minutes until the headloss value decreased from 2 m to 1 m. The following working period, during which the headloss increased from 1 m to 2 m, lasted approximately 30 days afterwards another washing step was necessary.

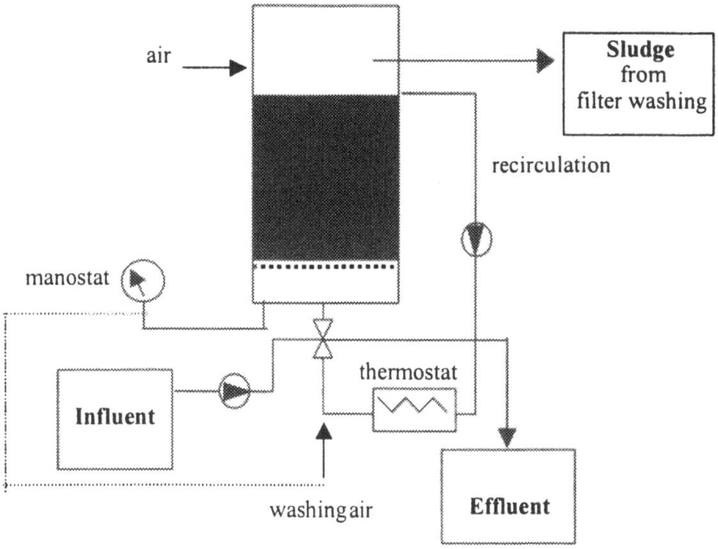

Figure 1: Sketch of the sequencing batch biofilm reactor used during the investigation.

During washing operations, the removed biomass was collected and measured as TSS and VSS in order to calculate the specific sludge production that resulted 0.05 kg TSS/kg $COD_{removed}$. The air-flow rate (100 L/h) was controlled by a flow meter, the temperature in the reactor was maintained at 20°C using a thermostat bath, and the pH, continuously monitored, was always in the range 7.0-7.5. Activated sludge from a local municipal wastewater treatment plant was used as inoculum.

Table 1: Lab plant SBBR features.

Diameter [cm]	23
Height [cm]	73
Geometric volume [L]	30
Fixed bed volume [L]	16
Packing media	KMT*
Initial bed porosity	0.75
Average working volume [L]	12
Surface upflow speed [m/h]	2.5÷3.0

* height: 7 mm; diameter: 8 mm; Dp = 2 mm; specific surface: 690 m^2/m^3; density: 0.95 g/cm^3

Wastewater

The reactor was fed with synthetic wastewater consisting of glucose (2,400 mgO_2/L), NH_4Cl (54.5 $mgNH_4$-N/L) and Na_2HPO_4 (18 $mgPO_4$-P/L).

Measurement methods and calculation procedures

Bed porosity ($m^3_{of\ empties}/m^3_{of\ bed}$) was determined according to the method developed by Jimenez et al. (1988). This method permits evaluation, by using a tracer (dextran blue), of the residence time in a submerged filter and then the calculation of bed porosity. Pressure losses were measured by a manostat at the bottom of the bed. Biomass concentration was evaluated in terms of dry weight, as TSS, on a representative bed volume. This measure lumps together granules and biomass attached to the carrier. The hydrodynamic shear forces were calculated by a procedure developed by the authors. According to this procedure (in Appendix A is reported the full description) shear force (τ) was calculated by following equation:

$$\tau = \frac{25\mu F}{D_p S_R}\frac{(1-\varepsilon)}{\varepsilon^2} + \frac{1{,}75\rho F^2}{S_R^2}\frac{1}{\varepsilon^2} \qquad (1)$$

where: μ is the absolute fluid viscosity (kg/m·s), ρ is the fluid density (kg/m^3), ε is the bed porosity ($m^3_{of\ empties}/m^3_{of\ bed}$), D_p is the equivalent diameter (m) of filling particles (carrier or granule), F is the recirculation flow rate (m^3/s), S_R is the section area (m^2),

The equation (1) allows the calculation of shear force after having measured the bed porosity and D_p, being all the other terms (i.e.: F, S_R, μ, ρ) known. The bed porosity was experimentally measured whereas D_p was calculated by measuring pressure loss (Δp) and applying the following equation:

$$\frac{(\Delta p)}{h}\frac{D_p S_R^2}{\rho F^2}\frac{\varepsilon^3}{(1-\varepsilon)} = 150\frac{S_R\mu(1-\varepsilon)}{\rho F D_p} + 1.75 \qquad (2)$$

Results and discussion

During the start-up period (about three months), biomass granulation was obtained according to the following steps. In particular, during the first days of the experimentation it was observed that a thin biofilm completely covered the carrier surface with an average biofilm concentration in the filter bed of about 7 gTSS/L_{bed} and a headloss lower than 10 cm. In the following days, an increase of biofilm thickness was recorded that accordingly led to an increase of both biomass concentration and headlosses up tp 20 g TSS/L and 1m respectively. Later on, the formation of biomass granules in external (i.e., pores produces by packing) and internal pores of the carrier took place. The figure 2 reports a picture of biomass before and after granulation process. At the end of start-up period, the headloss along the whole bed and the average biomass concentration was 2 m and 33 gTSS/L, respectively, and a first washing step was carried out for decreasing the headlosses from 2 to 1 m. On the basis of the above experimental observations, a schematic representation of how biomass granulation takes place during the SBBR start-up period has been outlined in figure 3, where four steps are sketched: *(1)* formation of a thin biofilm that fully covers the carrier; *(2)* biofilm thickness increase; *(3)* attached biofilm break-up with release of biofilm particles; *(4)* rearrangement of biofilm particles in smooth granules.

The generation of granular biomass can be also followed by observing the D_p profile, i.e. the equivalent diameter of the particles constituting the filling material of the bed during start-up period. At t = 0, i.e. in the absence of biomass growth, D_p is the equivalent diameter of the carrier, i.e. 2 mm (see table 1). Because of the progressive colonisation of the carrier (i.e. carrier covered by biofilm), D_p should increase according to the above equation (1).

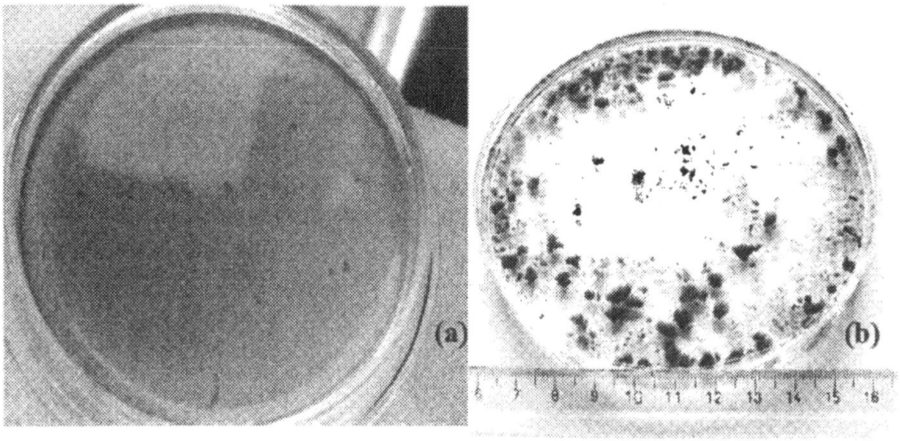

Figure 2: Biomass picture before (a) and after (b) granulation.

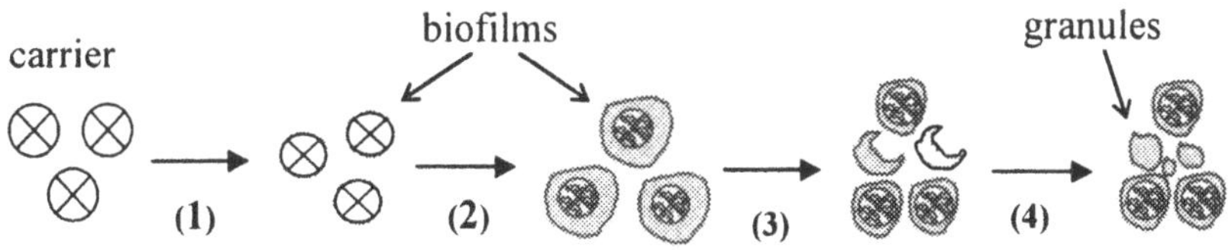

Figure 3: Sketch of the granules generation steps during the start-up period.

When granulation takes place, even the biomass granules act as filling particles. Therefore, D_p becomes the average equivalent diameter of all particles present in the bed (i.e., the carrier covered by biofilm and the biomass granules). By measuring pressure loss and bed porosity during the start up period and applying the above equation 2, the D_p value profile shown in figure 4 was obtained. Looking at this profile it is possible to observe that in the first 75 days D_p continuously increases reaching its maximum value, i.e. about 8 mm. At this point, i.e. when the carrier is covered by a very thick biofilm, D_p undergoes a sharp decrease from 8 mm to 1 mm. This latter very low D_p value can be justified only supposing that a large amount of the biomass attached to the carrier is detached forming particles most of which characterised by small sizes.

Inserting in equation 13 (see appendix A) the calculated D_p and the measured bed porosity values, it is possible to calculate the profile of specific area of the bed (m^2/m^3 of bed) shown in figure 4. Like the D_p trend, the specific area of the bed also has a sharp change after 75 days. In particular, it is possible to note that the biomass granulation leads to a specific surface increase of about six times (i.e. from 1,000 to about 6,000 m^2/m^3), and this accordingly causes an increase of biomass-liquid mass-transfer rate. As already said, the hydrodynamic shear force trend during the start-up period provides an explanatory key for the generation process of granular biomass. In fact, inserting in equation (1) the values of known parameters (i.e. F, S_R, μ, ρ), the following equation is obtained:

$$\tau = 1.74 \times 10^{-4} \frac{}{D_p} \frac{(1-\varepsilon)}{\varepsilon^2} + 0.0084 \frac{1}{\varepsilon^2} \quad (3)$$

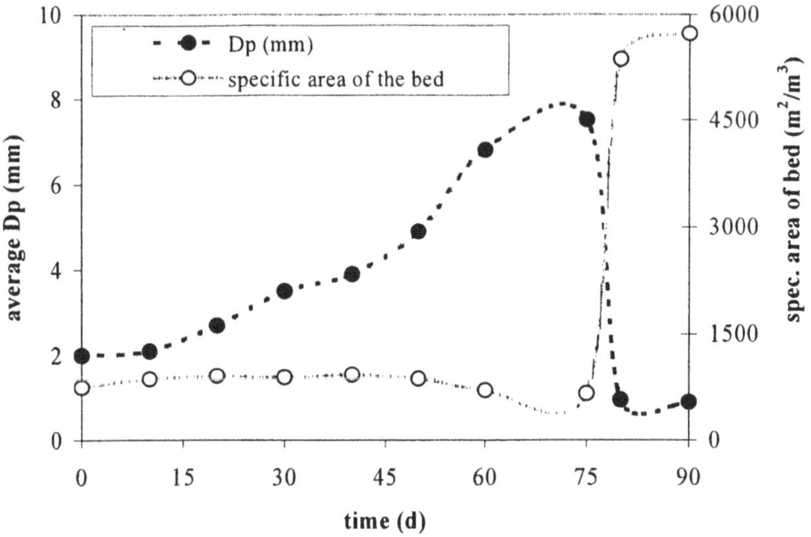

Figure 4: D_p and specific area of bed profiles during the start-up period.

Equation (3) expresses the hydrodynamic shear forces in terms of bed porosity and particle diameter (D_p), i.e. two parameters that continuously change during the start-up period because of progressive biomass growth and build up in the reactor. Inserting into equation (3) the measured bed porosity and the calculated D_p values, the hydrodynamic shear force profile, shown in figure 5, has been obtained. Such a profile quantitatively describes the hydrodynamic shear forces trend during the start-up period that includes the following four steps:

Step 1. *Formation of biofilm.* During this first step, i.e. at the beginning of the start up, the formation of a conventional thin biofilm take places that entirely covers the carrier surface.

Step 2. *Increase of the biofilm thickness.* During this step a biofilm thickness increase is recorded, as proved by the calculated Dp values (see figure 4). In addition, the profile in Figure 5 shows that the SBBR is characterised by rather low shear forces values (lower than 1 dyne/cm^2). Under such conditions, it is possible to assume that in this phase the biofilm detachment is primarily regulated by biofilm growth. Therefore, to avoid a premature sloughing events it is important to control the biofilm growth by regulating the substrate loading rate. In fact, if the start up period is too short, the premature biofilm detachment takes place and then granulation doesn't occur. Under the weak shear forces characterizing the step 2, the biofilm continues to grow increasing its thickness and assuming the porous structure characteristic of weak shear forces.

Step 3. *Break-up of biofilm.* The continuous increase of biofilm thickness causes the progressive increase of the shear forces. Such a situation has a negative effect on biomass stability. In fact, the biofilm grown with a porous structure characterized by a little adhesion strength is greatly affected by the increase of the applied detachment forces (i.e. shear forces) and, consequently, sloughing phenomena take place. The detachment of biofilm particles, in turn, causes a consequent quick D_p decrease (see figure 4). Furthermore, as the detached particles remain into the bed, the bed porosity does not change and the resulting D_p decrease causes the sharp hydrodynamic shear force increase shown in figure 5. The capacity of the bed to retain detached biofilm particles plays a decisive role on the shear forces increase and, as discussed later on, even on the granulation process. In fact, if the biofilm particles were expelled from the bed, the bed porosity would increase with the consequent reduction of the shear forces value and the return back to step1. The authors experimented other carriers with different features (i.e. with large pore volume) and observed that granulation didn't take place because the detached biofilm particles were expelled from the bed. The capacity of the bed to retain detached biomass particles can be ascribed to the specific features of the used carrier whose size and shape are such to generate a bed characterized by rather uniform internal and external small-sized pore volumes that are progressively reduced as a result of biomass growth.

Step 4. *Granule generation.* The high shear forces promote the rearrangement of the detached biofilm particles in smooth granules. In fact, the high shear forces reduce the number of biofilm particles protuberances that are continuously removed and, accordingly, the surface of biofilm particles becomes more and more smooth and less affected by shearing phenomena (van Loosdrecht *et al.*, 1995). Furthermore, during this phase, granule growth also take place as proved by the decrease of the measured bed porosity that leads to a further increase of shear forces.

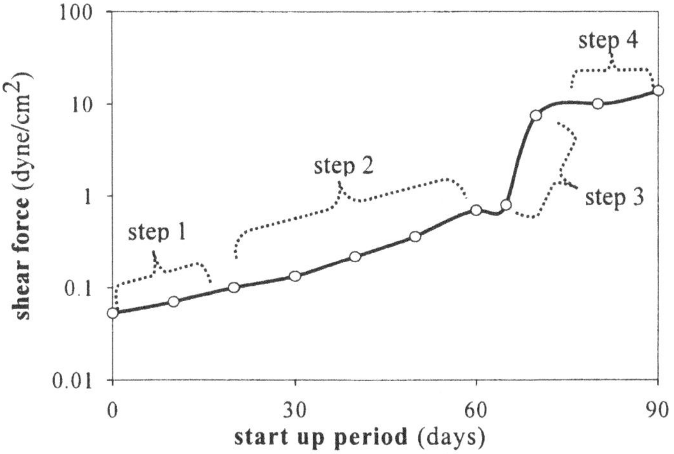

Figure 5: Hydrodynamic shear force profile during start-up period.

Conclusions

The main results obtained in experimental study specifically aimed at assessing a plausible mechanism for biomass granulation during the start up of a SBBR are the following:

- the granulation process takes place through four steps: (1) formation of a thin biofilm that fully covers the carrier; (2) increase of biofilm thickness; (3) break-up of attached biofilm with release of biofilm particles; (4) rearrangement of biofilm particles in smooth granules.

- the hydrodynamic shear force trend during the start-up period provides an explanatory key for the generation process of granular biomass.
- the bed features when sloughing events take place are crucial in the granules generation because they must guarantee the retention of detached biofilm particles.
- the granulation occurrence leads to an increase (about six times) of bed specific surface with consequent improvement of liquid mass-transfer rate.

Appendix A

Shear force and pressure loss (Δp) are linked by the following equation (force balance equation) (Foust *et al.*, 1980):

$$\tau \cdot A = \Delta p \cdot S \tag{4}$$

where S is the crossing area of the liquid and A is the friction surface.

For cylindrical pipes ($S=\pi D^2/4$; $A=\pi D h$) the following equation is obtained:

$$\tau = (\Delta p \cdot D)/4h \tag{5}$$

where h and D are pipe height and diameter, respectively.

For non cylindrical pipes, the equivalent diameter D_{eq} must be used (Foust *et al.*, 1980):

$$D_{eq} = 4\frac{S}{b} \tag{6}$$

where b is wet perimeter.

For a granular media bed that can be considered as an interconnected array of pipes, as b determination would be very difficult, it is necessary to modify the equation (6). Accordingly, multiplying and dividing the equation (6) by h, the following equation is obtained:

$$D_{eq} = 4\frac{S}{b} x \frac{h}{h} \tag{7}$$

i.e.:

$$D_{eq} = 4\frac{total\ volume\ of\ empties}{total\ surface\ area\ of\ granular\ media} \tag{8}$$

Multiplying and dividing equation (8) by bed volume, the following is obtained:

$$D_{eq} = 4\frac{\varepsilon}{S_{sb}} \tag{9}$$

where ε is bed porosity ($m^3$of empties/m^3 di bed), and S_{sb} is the surface area per unit volume of the bed (m^2/m^3 of bed).

Including equation (9) into (5) the following is obtained:

$$\tau = \frac{\Delta p}{h}\frac{\varepsilon}{S_{sb}} \tag{10}$$

this equation allows to calculate the shear forces providing that bed porosity, pressure loss and specific surface area (m^2/m^3 of bed) are known. In the experimentation presented in the paper, the first two parameters were determined experimentally, whereas the third one was calculated as follows:

converting the specific area per unit of the bed (i.e. S_{sb}) to the specific area per unit volume of particle (S_{sc}), it is necessary to multiply S_{sc} by $(1-\varepsilon)$:

$$S_{sb} = S_{sc}(1-\varepsilon) \tag{11}$$

S_{sc} ($m^2/m^3{}_{\text{of particle}}$) is the ratio between the surface, A_p (m^2), and volume, V_p (m^3), of the particle. For non spherical particles, the following equation is valid (Foust *et al.*, 1980):

$$D_p = 6\frac{V_p}{A_p} = \frac{6}{S_{sc}} \tag{12}$$

where D_p is the equivalent diameter of filling particles, i.e. the average diameter between all carriers covered by biofilm and granules.

Including equation (12) into (11) the following is obtained:

$$S_{sb} = \frac{6(1-\varepsilon)}{D_p} \tag{13}$$

D_p can be calculated by the following equation (Foust *et al.*, 1980):

$$\frac{(\Delta p)}{h}\frac{D_p S_R^2}{\rho F^2}\frac{\varepsilon^3}{(1-\varepsilon)} = 150\frac{S_R\mu(1-\varepsilon)}{\rho F D_p} + 1.75 \tag{2}$$

where: μ (kg/m·s) and ρ (kg/m^3) are viscosity and density of the fluid, respectively; F (m^3/s) is flow rate through a bed with a section area S_R (m^2), a height h and a porosity ε, consisting of random packed particles of diameter just D_p.

Therefore, equation (2) permits to calculate D_p because all the other terms in the equation are known or measurable parameters.

Including equation (13) and (2) into (10) the equation (1) is obtained:

$$\tau = \frac{25\mu F}{D_p S_R}\frac{(1-\varepsilon)}{\varepsilon^2} + \frac{1{,}75\rho F^2}{S_R^2}\frac{1}{\varepsilon^2} \tag{1}$$

References

Beun J.J., van Loosdrecht M.C.M., Heijnen J.J. (2002). Aerobic granulation in a sequencing batch airlift reactor. *Wat. Res.*, **34**, 702-712.

Di Iaconi C., Ramadori R., Lopez A. (2004). Preliminary biomass characterization in a sequencing batch biofilm reactor. *J. of Anal. and Environ. Ch.* (in press).

Etterer T., Wilderer P.A.(2001). Generation and properties of aerobic granular sludge. *Wat. Sci. & Technol.* **43**, 19-26.

Foust A.S., Wenzel L.A., Clump C.W., Maus L., Andersen L.B. (1980). Principles of unit operations. Wiley, New York.

Jimenez B., Noyola A., Capdeville B., Roustan M., Faup G. (1988). Dextran blue colorant as a reliable tracer in submerged filters, *Wat. Res.*, **22**, 1253-1257.

Liu Y. and Tay J.H. (2002). The essential role of hydrodynamic shear force in the formation of biofilm and granular sludge – Review. *Wat. Res.* **36**, 1653-1665.

McSwain B.S., Irvine R.L and Wilderer P.A. (2004). The influence of settling time on the formation of aerobic granules and extracellular polymeric substances. Proceeding of *3rd IWA Specialised Conference on Sequencing Batch Reactor Technology (SBR3)*, Noosa, Queensland, Australia, 22-26 February.

Morgenroth E., Sherden T., van Loosdrecht M.C.M., Heijnen J.J., Wilderer P.A. (1997). Aerobic granular sludge in a sequencing batch reactor. *Wat. Res.*, **31**(12), 3191-3194

Mulder R., Vereijken T.L.F.M., Frijters C.M.T.J., Vellinga S.H.J. (2001). Future perspectives in bioreactor development. *Wat. Sci. & Technol.* **44**(8), 27–32.

Pujol R., Canler J.P., Iwema A. (1992). Biological aerated filters: an attractive and alternative biological process. *Wat. Sci. & Tech.,* **26** (3/4), 693-702.

van Loosdrecht M.C.M., Eikelboom. D., Gjaltema A., Mulder A., Tijhuis L., Heijnen, J.J.(1995). Biofilm structures. *Wat. Sci. & Technol.*, **32**(8), 35-43.

Study on the Stability of Aerobic Granules in a SBAR-Effect of Superficial Upflow air Velocity and Carbon Source

Fang WANG, Yi-hui LIU, Feng-lin YANG*, Xing-wen ZHANG, Han-min ZHANG

School of Environmental Science and Technology, Dalian University of technology, No.2 Linggong road, Ganjingzi district, Dalian, 116023, P. R. China

Abstract The factors influencing the stability of aerobic granules were studied in three sequencing batch airlift reactors (SBAR). Superficial upflow air velocitie of 2.2 cm/s and 3.3 cm/s was supplied in the reactors when sucrose supplied as carbon source and aerobic granules were observed by light microscopy. The aerobic granules at superficial upflow air velocity of 3.3 cm/s were formed with smooth surface and good settling while those at superficial upflow air velocity of 2.2 cm/s were broken down for heavy growth of filamentous organisms. But when sludge loading was below 0.5 Kg COD/(Kg MLSS d), it was found that granular sludge also became unstable at superficial upflow air velocity of 3.3 cm/s with sucrose as carbon source because of heavy growth of filamentous organisms in the reactor. But mixed carbon source, such as sucrose and sodium acetate, improved the stability of aerobic granules distinctly.

Keywords Aerobic granulation, carbon source, sludge loading rate, stability, superficial upflow air velocity

Introduction

Sludge granulation is the results of a microbial self-immobilization process, and it is considered as biofilm formation without supporting carrier material. Aerobic granular sludge has some advantages over conventional bioflocs, such as good settling ability, high biomass retention and higher ratio of F/M (Peng et al., 1999; Beun et al., 2000; Ettere et al., 2001). Therefore, study of aerobic granulation is becoming popular in the field of wastewater treatment.

Studies (Jang et al., 2003; Tay et al., 2002; Toh et al., 2003) on aerobic granules have achieved many results. Studies indicate that the microbial structure and properties have a close relation with substrate N/C ratio (Liu et al., 2003) and operation conditions (Moy et al., 2002). Therefore, in this study, an investigation was performed onto the effect of superficial upflow air velocity on the formation of aerobic granules while sucrose acted as sole carbon source. Further more was the influence of different carbon sources on the aerobic granulation investigated. It is expected that the experimental findings will provide useful information or guides for improving the stability of aerobic granules.

Materials and methods

During the study, the experimental set-up included three sequencing batch airlift reactors R1, R2 and R3, each with the same geometrical configuration. The reactor had a working volume of 3.4 L (Figure 1). The diameter of the down-comer was 80 mm. The riser was 900 mm in height, had an internal diameter of 40 mm, and was positioned at a distance of 20 mm from the bottom of the reactor. Effluent was withdrawn at 400 mm height from the bottom of the reactor (equal to a VER of 55%). A fine bubble aerator at the bottom of the reactor introduced air into the reactor.

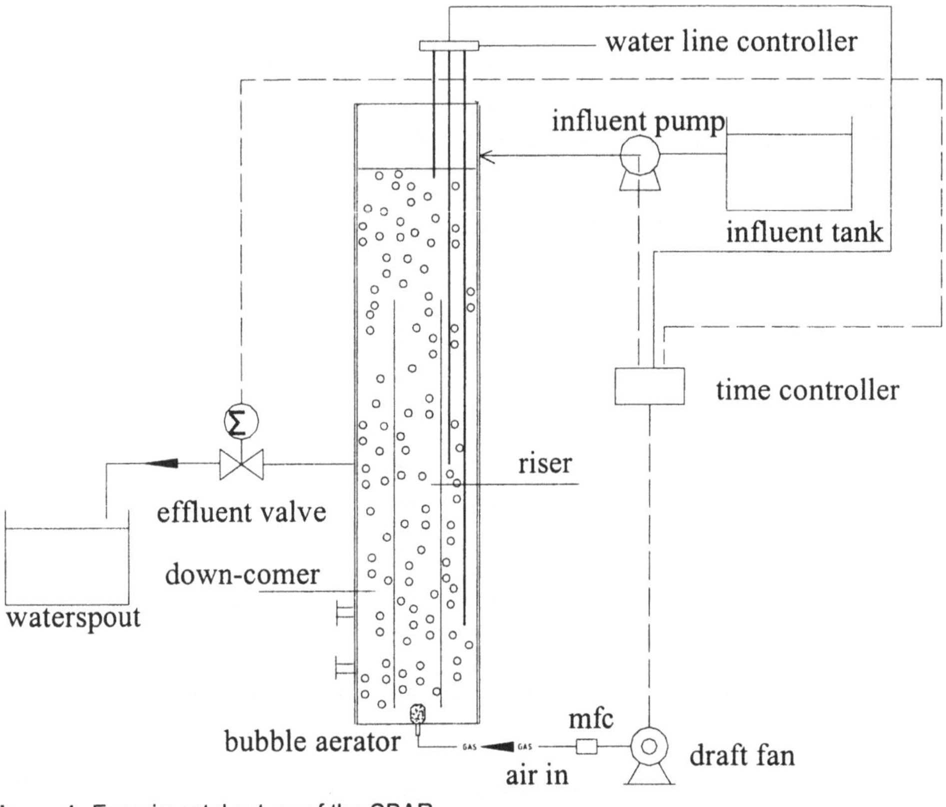

Figure 1: Experimental set-up of the SBAR

R1 and R2 were supplied with a superficial upflow air velocity of 2.2 cm/s and 3.3 cm/s respectively while sucrose was supplied as sole carbon source in both reactors. A mixture of sodium acetate and sucrose was used as carbon source in R3 at a superficial upflow air velocity of 3.3 cm/s. Composition of the synthetic wastewater in R1 and R2 is shown as follows: sucrose, 700 mg L^{-1}; NH_4Cl, 190 mg L^{-1}; KH_2PO_4, 20 mg L^{-1}; $CaCl_2{\cdot}2H_2O$, 31 mg L^{-1}; $MgSO_4{\cdot}7H_2O$, 94 mg L^{-1}; $FeSO_4{\cdot}7H_2O$, 22 mg L^{-1}, and 1mL L^{-1} micronutrients. In R3, the composition is: sucrose, 300 mg L^{-1}; sodium acetate, 670mg L^{-1}; NH_4Cl, 190 mg L^{-1}; KH_2PO_4, 20 mg L^{-1}; $CaCl_2{\cdot}2H_2O$, 31 mg L^{-1}; $MgSO_4{\cdot}7H_2O$, 94 mg L^{-1}; $FeSO_4{\cdot}7H_2O$, 22 mg L^{-1}, and 1mL L^{-1} micronutrients. The micronutrient solution contained (mg L^{-1}): EDTA, 15000; H_3BO_4, 14; $ZnSO_4{\cdot}7H_2O$, 430; $CuSO_4{\cdot}5H_2O$, 250; $MnCl_2{\cdot}4H_2O$, 990; $NaMoO_4{\cdot}2H_2O$, 220; $Na_2SeO_3{\cdot}5H_2O$, 210; $AlCl_3$, 50; $CoCl_2$, 50; and $NiCl_2$, 199(Moy et al., 2002).

All reactors were inoculated with 300 ml of activated sludge from a municipal wastewater treatment plant. The temperature was controlled at 22 ± 2 °C and the pH of influent was maintained about 7.5 ± 0.5 using $NaHCO_3$. All reactors were operated in sequencing batch mode at a total cycle duration time of 3 h: 10 min of influent filling, 163 min of aeration, 5 min of settling and 1 min of effluent discharging.

Influent and effluent from all reactors were analyzed for COD_{tol} and the sludge for mixed liquor suspended solids (MLSS), mixed liquor volatile suspended solids (MLVSS) and sludge volume index (SVI) according to standard methods (APHA, 1998). Dissolved oxygen was determined using an oxygen electrode of 55-YSI.

The morphology of the sludge was characterized using light microscopy. The microbial morphology of granules was examined in more detail with a scanning electron microscope.

Results and discussion

The influence of superficial upflow air velocity on the formation of aerobic granules

R1 was operated at the superficial upflow air velocity of 2.2 cm/s. After five days of reactor running, SVI of the sludge decreased from 103.5 ml/g to 47.2 ml/g, and there were small granules present in reactor; after ten days of reactor running, it was found that filamentous organisms appeared in the reactor and small granules with fluffy shape present while SVI increased to 68.7 ml/g. From then on, heavy growth of filamentous organisms was observed (shown in figure 2) and SVI increased above 170 mg/l. As a result, operation of R1 was breakdown because of bad settling of sludge and washout of biomass from the reactor.

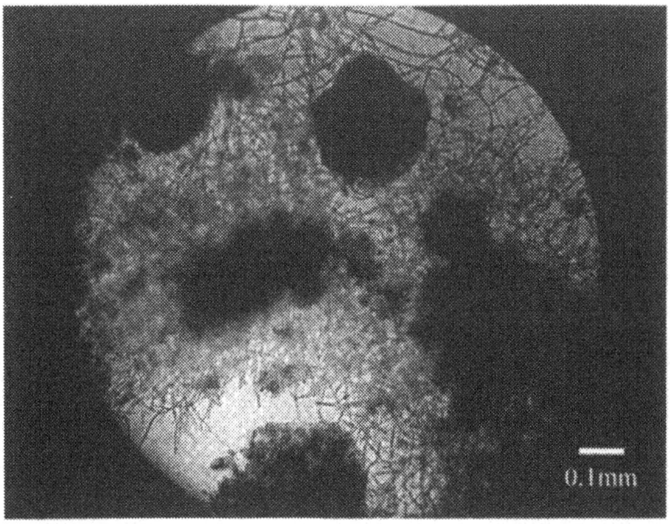

Figure 2: Microphotographs of granular sludge at superficial upflow air velocity of 2.2 cm/s (day 15 of reactor running)

However, in R2 after running for six days, many small granules appeared and SVI decreased to 26.5 ml/g. After fifteen days, granules with a clear round outer shape were prevailed and floc sludge was little while there were only few filamentous organisms present (shown in figure 3). Subsequently, the SVI of the granules kept around 18.6 ml/g with a smooth surface.

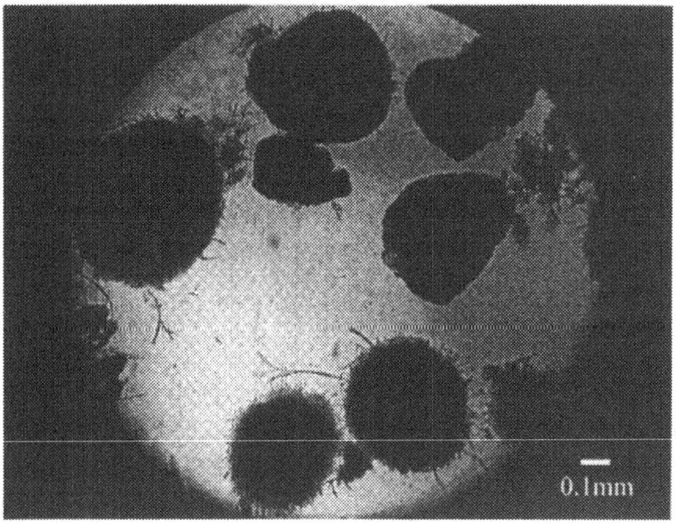

Figure 3: Microphotographs of granular sludge at superficial upflow air velocity of 3.3 cm/s (day 15 of reactor running)

Aerobic granules formed under mixed carbon source

During the running of R3, sodium acetate and sucrose were mixed as carbon source in the influent of the reactor. After fifteen days of R3 operation, small granules present in the reactor and SVI decreased to 31.6 ml/g while there was no filamentous organisms appeared (shown in figure 4). After then, the granulation of sludge still got well and there were no filamentous organisms existed in the whole operation.

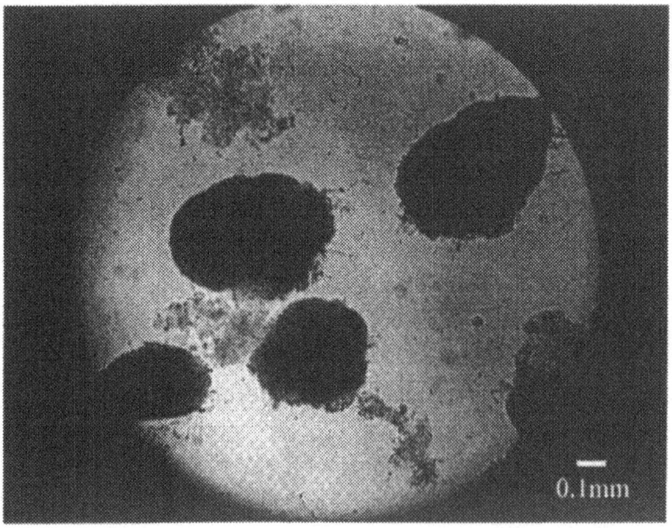

Figure 4: Microphotographs of granules under sucrose and sodium acetate as mixed carbon source (day 15 of reactor running)

The effect of sludge loading rate on the stability of aerobic granules at a superficial upflow air velocity of 3.3 cm/s

During the granulation of sludge, the organic loading rates in R2 and R3 were maintained to 3.29 kg COD / (m^3d) and MLSS was around to 3500 mg/L, as a result, the sludge loading rate (SLR) was 0.9 kg COD/(Kg MLSS d). After the formation of aerobic granules with smooth surface and good settling, SLR was decreased with the increase of MLSS. It was found that the settling ability of granules in R2 became worse and worse while that of granules in R3 still kept well when SLR was below 0.5 kg COD/ (Kg MLSS d) in both reactors. The changes of SVI with SLR were shown in figure 5. And it was found that the filamentous organisms shot out from the surface of granular sludge in R2 (shown in figure 6 a), but still no filamentous organisms present in R3 (shown in figure 6 b) when both reactors were operated 35 days long. After that, with the increase of filamentous organisms in R2, biomass was run out from the reactor during stage of effluent withdrawal and operation of R2 was broken down.

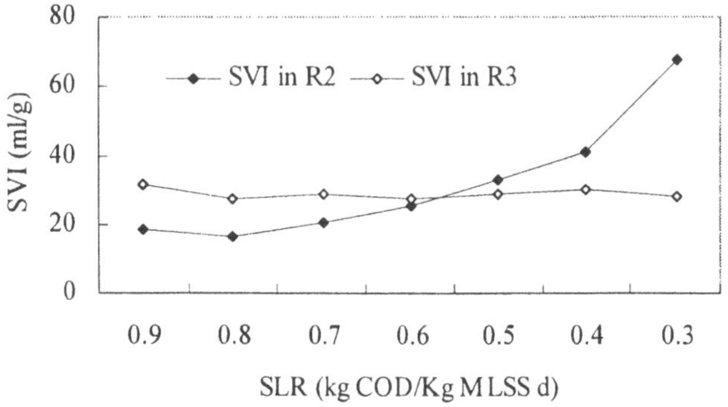

Figure 5: Changes of SVI with SLR in R2 and R3

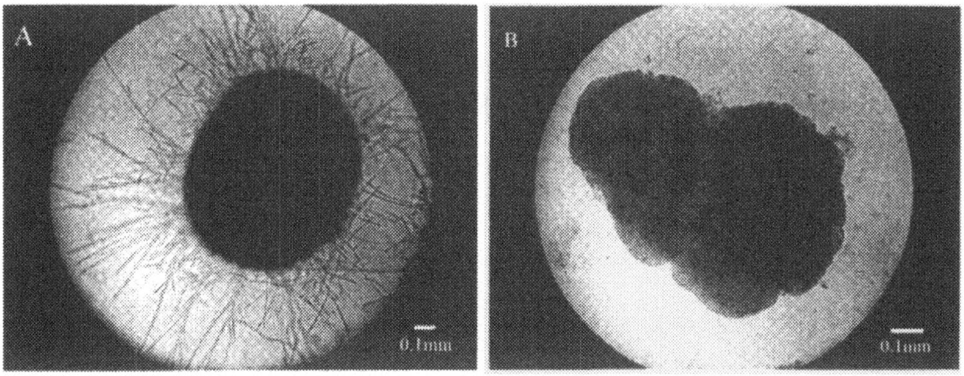

Figure 6: Microphotographs of granular sludge under SLR of 0.4 kg COD/ (Kg MLSS d) (a) in R2; (b) in R3

Further, the detail analysis on the surface morphology of aerobic granules respectively in R2 and R3 when SLR decreased to 0.4 Kg COD/ (Kg MLSS d) was carried out by a scanning electron microscopy, as shown in figure 7.

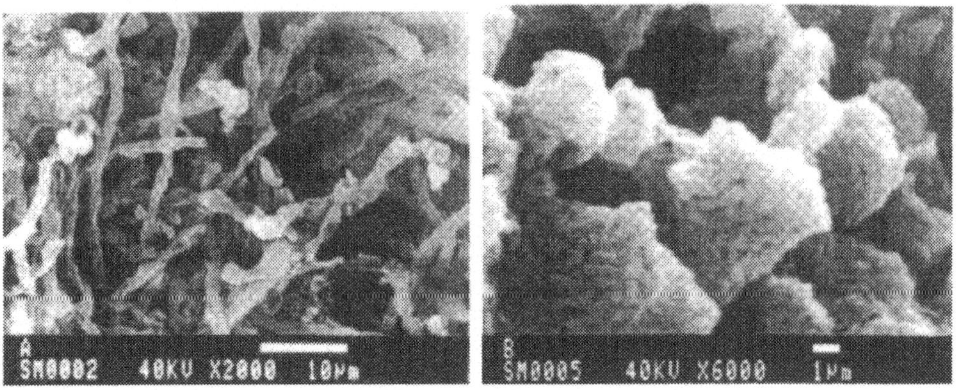

Figure 7: High magnification of aerobic granular sludge surface by SEM (a) sucrose as sole carbon source; (b) sucrose and sodium acetate as mixed carbon source

It was found that surface of the aerobic granules formed under sucrose was mainly composed of filamentous organisms, but surface of the aerobic granules formed under sucrose and sodium acetate was no filamentous organisms but zoogloea bacteria.

Analysis on the instability of aerobic granules with sucrose as sole carbon source

There are many researches focused on the changes in granule structure which is depending on the reactor operational conditions and composition of their wastewater (Tay et al., 2003a, 2003b; Yang et al., 2004). From our results, it was found that, when sucrose acted as sole carbon source in R1 and R2, the crucial reason for instability of aerobic granules in the reactors was the heavy growth of filamentous organisms. Although it was easy to cause heavy growth of filamentous organisms with sucrose as sole carbon source, the sludge granulated well and no filamentous organisms appeared when fed with a mixed carbon source in R3.Wang (1999) showed that, a high carbohydrate concentration at DO-concentration around 1.0~2.0 mg/L in the mixed liquor restricts the growth of zoogloea bacteria as a result of lacking oxygen, at the same time it favours the growth of filamentous organisms which can grow at low dissolved oxygen. DO in one cycle time of the reactors operated 10 days is shown in figure 8.

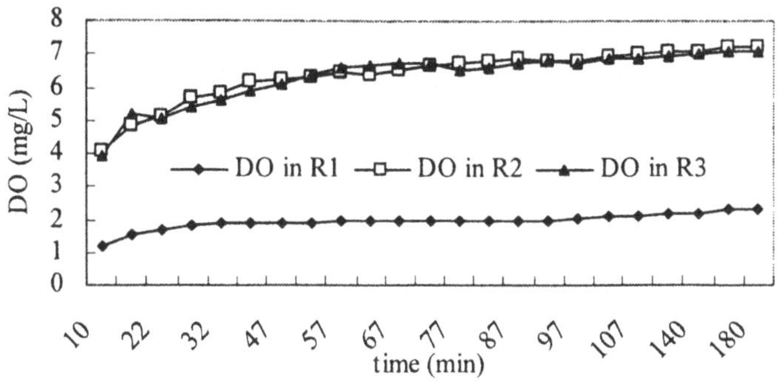

Figure 8: Dissolved oxygen of mix liquor in one cycle time of the reactors

As shown, DO in the mixed liquor was around 2 mg/L at the superficial upflow air velocity of 2.2 cm/s in R1. Therefore, the growth of filamentous organisms was too heavy during the process of sludge granulation for low DO in mix liquor of R1and finally caused aerobic granules not to form with smooth surface and good settling. At the superficial upflow air velocity of 3.3 cm/s in R2, DO in the mixed liquor was high which led to the development of aerobic granules with smooth shape and few filamentous organisms in the reactor. But, after the formation of granules and with decrease of SLR in R2, filamentous organisms also became dominant. Wang (1995) and Peng (1996) indicated that the kinetic growth constant of filamentous organisms is smaller than that of zoogloea, but under low SLR, zoogloea bacteria cann't gain sufficient substrate while filamentous organisms can develop long filament in order to increase its surface area and obtain as much substrate as possible, so growth of filamentous organisms is faster than that of zoogloea, as a result, filamentous organisms become dominant in reactor of R2. Another, it is thought (Li et al., 2002) that filamentous organisms are able to utilize saccharide directly as energy source and reproduce easily when saccharide is dominant, so filamentous organisms was present in R2 when sucrose was supplied as sole carbon source when SLR was below 0.5 Kg COD/ (Kg MLSS d), but no filamentous organisms present in R3 because saccharide content was comparatively small when supplied with mixed carbon source.

Conclusions

This study demonstrated the behaviour of aerobic granular sludge during the startup period of three different SBAR-systems, and the reasons for the instability of aerobic granules were discussed. It is concluded as follows:

DO in the mixed liquor was around 2mg/L at the superficial upflow air velocity of 2.2 cm/s but was high at the superficial upflow air velocity of 3.3 cm/s, so there was less filamentous organisms at the superficial upflow air velocity of 3.3 cm/s than at the superficial upflow air velocity of 2.2 cm/s when SLR is above 0.5 Kg COD/ (Kg MLSS d) for the reason that a high carbohydrate concentration at low DO-concentration in the mixed liquor restricts the growth of zoogloea bacteria but favours the growth of filamentous organisms which can grow at low dissolved oxygen..

As filamentous organisms grows faster than zoogloea under low SLR, granular sludge is unstable for filamentous organisms are dominant with sucrose as sole carbon source when SLR is below 0.5 Kg COD/(Kg MLSS d). Although filamentous organisms can utilize saccharide directly as energy source and reproduce easily, no filamentous organisms present when supplied with mixed carbon source.

Supplying sucrose and sodium acetate as mixed carbon source is a feasible way to control the heavy growth of filamentous organisms and resolve the question of poor stability of aerobic granules.

References

APHA. (1998) Standard methods for Examination of water and wastewater, 20th ed., American Public Health Association, Washington, DC

Beun JJ, Ban Loosdrecht MCM, Heijnen JJ. (2000) Aerobic granulation. Wat Sci Tech, 41(4-5):41-48

Etterer T, Wilderer PA (2000) Generation and properties of aerobic granular sludge Wat Sci Tech, 43(7):19-26

Jang Am, Yoon Young-hua, Kim InS, K Kim wang-Soo, Bishop PL. (2003) Characterization and evaluation of aerobic granules in SBR. J biotech, 105(1/2): 71-82

Li Jun, Yang Xiushan, Peng Yongzhen. (2002) Microorganism and wastewater treatment engineering: Chemical Technology publishing company, Beijing

Liu Y, Yang SF, Tay JH. (2003) Elemental compositions and characteristics of aerobic granules cultivated at different substrate N/C ratio. Appl microbio biotech., 61(2):556-561

Moy BYP, Tay JH, Toh SK, Liu Y, Tay STL. (2002) High organic loading influence influences the physical characteristics of aerobic sludge granules. Lett Appl Micro 34: 407-412

Tay JH, Ivanov V, Pan S, Tay STL. (2002) Specific layer in aerobically grown microbial granules. Lett appl micro,34(4):254-257

Tay JH, Pan S, Tay STL, Ivanov V, Liu Y (2003)The effect of organic loading rate on the aerobic granulation: the development of shear force theory Wat Sci &Tech, 47(11): 235-240

Tay JH, Tay STL, Ivanov V, Pan S, Jiang HL, Liu QS (2003) Biomass and porosity profiles in microbial granules used for aerobic wastewater treatment. Lett Appl Microb, 36, 297-301

Toh SK, Tay JH, Moy BYP, Ivanov V. (2003) Size effect on the physical characteristics of the aerobic granule in a SBR.Appl micro biotech, 60(6):687-695

Peng DC, Bernet N, Jean-philippe D, Rena M. (1999) Aerobic granular sludge –a case report. Wat Res, 33(3): 890-893

Peng Yongzhen. (1996) Relationship between sludge age and sludge settling. China water&wastewater, 12(4):24-27

Wang Fuzhen. (1995) Sludge bulking problem and sequencing batch reactor. China Environmental Science, 15(2):131-135

Wang Kaijun. (1999) Exoriment of controlling sludge bulking under high orangic loading. Water&wastewater Engineering, China, 25(11): 30-34

Yang SF, Liu QS, Tay JH, Liu Y(2004) Growth kinetics of aerobic granules developed in sequencing batch reactors. Lett Appl Microb, 38(2):106-112,

Microstructural optimization of wastewater treatment by aerobic granular sludge

V. Ivanov, J.-H. Tay, Q.-S. Liu, X.-H. Wang, Z.-W. Wang, B. A. M. Maszenan S. Yi, W.-Q. Zhuang, Y.-Q. Liu, S. Pan, and S.T.-L. Tay

Division of Environmental and Water Resources Engineering, School of Civil and Environmental Engineering, Nanyang Technological University, 50 Nanyang Avenue, Singapore 639798 (E-mail: cvivanov@ntu.edu.sg, cjhtay@ntu.edu.sg, cqsliu@ntu.edu.sg, wang0048@ntu.edu.sg, PG04508928@ntu.edu.sg, P147047247@ntu.edu.sg, cmaszenan@ntu.edu.sg, P144766972@ntu.edu.sg, CYQLiu @ntu.edu.sg, CSPan@ntu.edu.sg, ctltay@ntu.edu.sg)

Abstract Microbial granules used in aerobic treatment of wastewater have diverse microbial communities, complex spatial structures and coordinated physiological functions. The aim of this study was to investigate how the structures of microbial granules can lead to optimal engineering parameters for wastewater treatment. A variety of radial (perpendicular to the granule surface) aggregates and concentric layers were detected in the granules by confocal laser scanning microscopy (CLSM) using fluorescence in situ hybridization (FISH) with oligonucleotide probes, specific fluorochromes and fluorescent microspheres. Granules typically consisted of an external layer of obligate aerobic microorganisms, followed by facultative anaerobic, then obligate anaerobic bacteria (at a depth of 800 mm below the granule surface), and finally a core of dead and lysed cells. The presence of anaerobic bacteria can disrupt wastewater treatment by aerobic granules due to the production of acids and gases that cause floating of the granules. The optimal diameter of the aerobic granule is less than 1.7 mm considering absence of the layer of obligate anaerobic bacteria or less than 0.5 mm considering that the whole granule should have a porous biomass-filled matrix. Time of granule formation from the flocs of activated sludge was from 8 to 14 days but it can be diminished to 2 days if selected bacterial strains with enhanced self-aggregation will be used instead of activated sludge.

Keywords Microbial granule, structure, wastewater treatment, CLSM, oligonucleotide probe

Introduction

Aerobically grown microbial granules are actively investigated as bioagents for the biological treatment of wastewater. These microbial aggregates have a short settling time and the ability to treat high strength wastewater (Beun et al., 1999, 2000, 2002; Etterer and Wilderer, 2001; Morgenroth et al., 1997; Moy et al., 2002; Peng et al., 1999; Tay et al., 2001a, b, 2003a, b; Toh et al., 2002; Zhu and Wilderer, 2003). However, due to the dense aggregation of cells, the rate of mass transfer of nutrients and metabolites between bulk medium and granular matrix may not be sufficient to ensure normal cell metabolism in the granule interior.

The concentration of dissolved oxygen can drop to zero at some depth below the granule surface. This depth depends on the specific rate of oxygen consumption and also on the porosity and extent of channel structures in the granules. The typical depth of the aerobic zone in a thick microbial biofilm in the presence of aeration is between 50 to 200 mm (Villaverde and Fernandez-Polanco, 1999; Gieseke et al., 2001). It was also demonstrated earlier that obligate anaerobic bacteria can grow in the interior of aerobically grown granules (Tay et al., 2002a, b; Tay et al., 2003a). Hypothetically, this can be accompanied by the production of fermentation gases and organic acids, which can deteriorate or even destroy the granules. Therefore, the size of the granules must not exceed a threshold value to ensure long-term stability and high metabolic activity.

Another structural property of microbial granules related to their bioengineering functions is the arrangement of granule components as radial sub-aggregates, spherical sub-granules and concentric layers. These components can hypothetically affect the formation, stability and activity of the granules. An important engineering aspect of aerobic wastewater treatment using microbial granules is the ratio of the rates of granule growth and decay, which can also be determined from structural investigations. The aims of this research were therefore to study the structure of microbial granules and to apply this knowledge to optimize the engineering properties of the granules.

Materials and Methods

The microbial granules were produced in a column sequencing batch reactor with a medium containing glucose (for all experiments) and ethanol or acetate (for the experiments on biomass distribution in the granules), as the main source of carbon, as described earlier (Tay et al., 2001a, b). The flocs retention time and mean hydraulic retention time in SBR were 8 h. There were 6 cycles of effluent with flocs removal a day with discharge of 50% of SBR working volume after 2-5 minutes of granules settling. The diameter of the studied granules varied from 450 mm to 3000 mm. The granules and flocs for fluorescence and VSS measurements were separated after 2-5 min settling. To study distribution of microorganisms in the granules by fluorescence in situ hybridization (FISH), granules with an average diameter of 2.2 mm were selected manually using a pipette with an attached tip having an orifice of 2.2 mm diameter. Bacto1080 and Nsm156 probes were applied using the FISH procedure. The exopolysaccharide (EPS) matrix in the granule was detected with a FITC-labeled lectin (ConA-FITC) from Canavalia ensiformis. The LIVE/DEAD® BacLight Bacterial Viability Kit (Molecular Probes, OR, USA) was used to evaluate quantity of dead and viable biomass. Intensity of green fluorescence of biomass stained with SYTO™ from this kit correlated with the number of ribosomes in cells, which depends on specific growth rate. Therefore, staining with SYTO™ was used for the detection of the active biomass layer in the granule. TetraSpec Fluorescent Microsphere Standards (Molecular Probes, OR, USA) detected channels and pores with diameters greater than 0.1 mm. All were visualized with Fluoview300 confocal laser scanning microscope (CLSM) (Olympus, Japan) as described previously (Tay et al., 2002 a, b). Cloning and sequencing of the 16S rRNA genes of the bacteria in the granules and phylogenetic analyses of the cloned sequences were performed as described previously (Tay et al., 2002b). Physiological properties of the operational taxonomic units (OTUs) such as relation to oxygen were inferred from their phylogenetic identification. A CY5-labeled Ent1432 probe with the sequence 5¢-CTTTTGCAACCCACT-3¢ (Sghir et al., 2000) and with Tm of 45°C was used to detect enterobacteria. Final concentration of the probe before hybridization was 70 pmol mL-1. Samples were incubated for 4 h at 46°C and washed after hybridization with 1 mL of washing buffer (hybridization buffer without formamide) for 2 h at temperature 46°C.

To study the formation of radial microaggregates of nitrifying bacteria, an ammonia gradient was formed in a culture tube filled with Noble Agar (Difco™, BD, Franklin Lakes, NJ, USA). The deterioration of the granules was studied by labelling cells with 1 μg L-1 of fluorescent lipophilic tracer DiIC18(3) (Molecular Probes, OR, USA). The tracer was readily taken up by the cells. The in-solution concentration of the tracer one day after its introduction into the reactor was less than 1% of the concentration detected in a suspension of particles produced by disintegrating granules in a 2 mL tube with phosphate-buffered saline (PBS) using a Mini-Beadbeater (Biospec Products, Inc., Bartlesville, OK, USA) for 100 sec at 500 rpm. The amount of fluorescence due to the lipophilic stain was determined using a Luminescence Spectrometer LS-50B (Perkin-Elmer, Boston, MA 02118, USA). Background due to autofluorescence of biomass was excluded from the reported values. Aggregation index (AI) was determined as relative decrease of optical density (OD) of suspension at 660 nm after 2 min of centrifugation at 650×g. AI was expressed as percent of initial OD and reflected the percentage of cells precipitated as aggregates.

Results and discussion

The components of a granule

The matrix of aerobically grown microbial granules is not homogenous. There are different microaggregates inside the granules. They may be arranged randomly, in a radial direction (Fig.1a), or as concentric layers (Fig.1c). Larger granules may also result from merging of smaller granules. The formation of radial aggregates of nitrifying bacteria (Fig. 1b), in a direction that is normal to the granule surface, is probably driven by the co-existence of steep oxygen gradient and reverse ammonia gradient created by release of ammonia from the central core of granule where biomass is lysed. This hypothesis was examined in independent experiments on the growth of enrichment cultures of nitrifying bacteria in Noble Agar where oxygen was supplied through the agar surface but where ammonia was supplied from the bottom of the agar layer. The nitrifying bacteria formed two layers. The first layer was a uniform biofilm but the second layer contained aggregates of nitrifying bacteria aligned normal to the agar surface (Fig.1b). The decreasing widths of these nitrifying aggregates probably reflect the dependence of growth rate on the available concentration of dissolved oxygen and ammonia.

The occurrence of concentric layers in granules was demonstrated using CLSM after staining by specific fluorochromes or FISH with specific oligonucleotide probes. The description of the layers is given in Table 1. Considering a microbial granule as a sphere with a diameter of 2.4 mm, the volumes of different zones can be calculated and compared with the experimental microbiological diversity of the granules (Table 2). There was a statistically reliable correlation between the calculated volumes occupied by aerobic, facultative anaerobic and anaerobic bacteria and the experimentally determined percentages of aerobic, facultative anaerobic and anaerobic bacteria isolated from the granules.

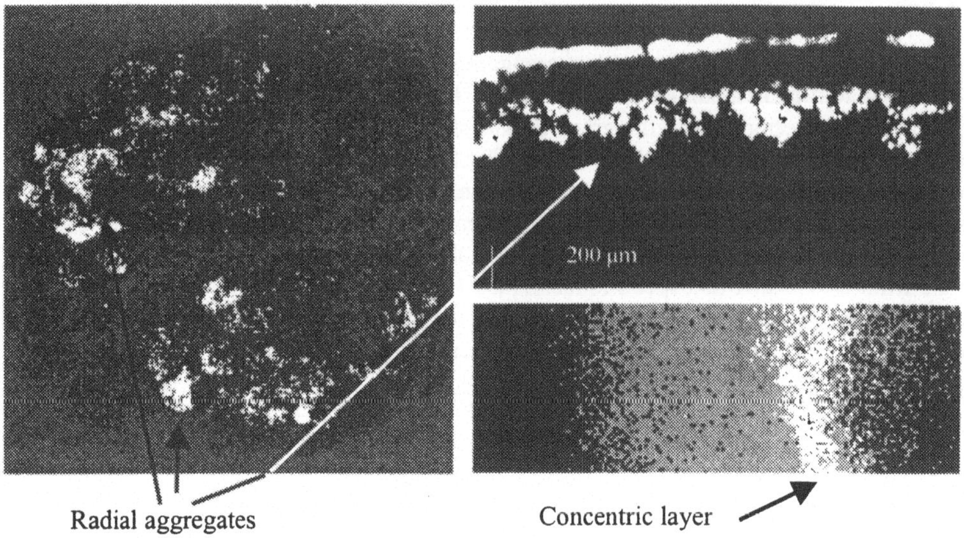

Figure 1: Different microaggregates of cell aggregates. **Figure 1a:** Radially arranged microaggregates of ammonia-oxidizing bacteria (bright structures) in a 3D image produced by CLSM of the granule. **Figure 1b:** Biofilm of nitrifying bacteria on the surface of Noble Agar with oxygen supply through the agar surface and ammonia supply from the agar bottom; one layer was a uniform biofilm but another one contained aggregates of nitrifying bacteria arranged perpendicular to the agar surface. **Figure 1c:** concentric layer of Bacteroides spp.

Table 1: Descriptions of layers in aerobically grown microbial granules grown in a column sequencing batch reactor with a medium containing ethanol or acetate (Tay et al., 2001a, b).

Layer	Average depth of layer from surface of granule and average thickness	Assumed function in the granule
Aerobic ammonia-oxidizing bacteria	70 μm (depth); 30 μm (thickness)	It reflects the depth of oxygen diffusion into granule
Facultative anaerobic enterobacteria	Concentration increased to maximum at a depth of 450 μm and remained stable at depths from 450 to 850 μm	Bacteria perform both aerobic and anaerobic processes
Obligate anaerobic bacteria *Bacteroides spp.*	850 μm (depth); 150 μm (thickness)	It reflects the presence anaerobic zone in granule
Channels and pores by penetration of 0.1 μm microspheres	Depth linearly depends on granule diameter by equation 1.	Deeper diffusion of nutrients
Layer of active biomass	Thickness linearly depends on granule diameter	All bioactivities of the granule are concentrated in this layer
Polysaccharides	Low content to a depth of 500 μm, reaching a maximum at 650 μm. Stable but low content at depth from 800 to 1200 μm.	It can decrease diffusion of nutrients into granule hrough the channels
Core of dying cells in the centre of granule	Depth was 1000 μm. Diameter of this inner core depended on granule diameter.	Supply of monomers and ammonia from this zone

Optimization of granule size

The concentric layers were typically arranged in sequence as obligate aerobic bacteria, facultative anaerobic bacteria, obligate anaerobic bacteria, and finally a core of dead and lysed cells. The presence of anaerobic bacteria can potentially diminish the stability of the granules due to the production of acids and gases from fermentation. Another negative effect of anaerobic bacteria on the wastewater treatment process is the occurrence of floating granules, which could occur if anaerobic bacteria are allowed to incubate in medium containing nitrate accumulated due to nitrification (Fig.2). There were anaerobic conditions in the layer of settled granules. Therefore, floating of the granules was probably due to gas production during denitrification, similar to the floatation of denitirifying granules (Etchebehere et al., 2002). This potential floating of the microbial granules in case of high organic or nitrate load leading to the production of gases in anaerobic zone of the granule can deteriorate wastewater treatment.

To avoid formation of anaerobic layer and core and possible deterioration of wastewater treatment the aerobic granules should have a diameter that is less than twice the distance from the granule surface to the anaerobic layer. This minimal distance is 0.85 mm (Table 2). Therefore, diameter of the granules without anaerobic layer and core should be less than 1.7 mm.

Table 2: Average geometric and biological parameters of 2.4 mm spherical granule grown in a column sequencing batch reactor with a medium containing ethanol or acetate (Tay et al., 2001a, b).

Layer or zone in the granule	Geometric parameters	Volume, mm^3	% of total volume of the granule	% of related bacterial clones isolated from the granules
Aerobic microorganisms in porous layer	0.55 mm below granule surface	6.09	84.1	69 ± 7%
Facultative anaerobic microorganisms	Between 0.55 mm and 0.85 mm below granule surface	0.97	13.4	9 ± 7%
Obligate anaerobic microorganisms (*Bacteroides spp.*)	Between 0.85 mm and 1.0 mm	0.15	2.0	2.1 %
Central core of dead and lysed cells	Depth is 1 mm. Diameter is 0.4 mm	0.03	0.5	

Low concentration of substrates and high concentration of metabolites in microbial aggregates may be overcome if channels and pores are present which allow advective mass transport. The mass transfer in channels and pores has been previously observed in aerobic biofilms (Massol-Deya et al., 1995). The aerobic granules may also contain channels and pores (Table 1). However, plugging of the channels and pores by polysaccharides, which are producing in the granules (Table 1), will likely promote the formation of the anaerobic layer and cell death in the central core of the granule. Therefore, another approach of size optimization is based on the assumption that whole granules should have a porous biomass-filled matrix without a core filled by dead and lysed cells. Depth and thickness (Hl) of the layer of porous biomass linearly correlated with granule diameter (Dg) by equation:

$$Hl = 0.15\ mm + 0.2Dg \qquad (1)$$

The optimal size of the aerobically grown microbial granule (Dc) may be calculated from Eq. 1 using the condition that 2Hl = Dg, which means that whole granule with diameter less than or equal to Dc is consisting entirely of a porous matrix. The value of Dc calculated from Eq.1 for this condition is 0.5 mm.

Physiological parameters such as specific COD removal or oxygen uptake rate cannot be used for conclusion on optimal diameter of the granules because increase of granule size diminishes the TOC/COD removal rate per 1 g of VSS of the granules (Toh et al., 2003). Conditions in the sequential aerobic sludge blanket reactor can be imposed to enhance wastewater treatment efficiency by selecting for microbial granules with a diameter smaller than the critical diameter.

A B C D E

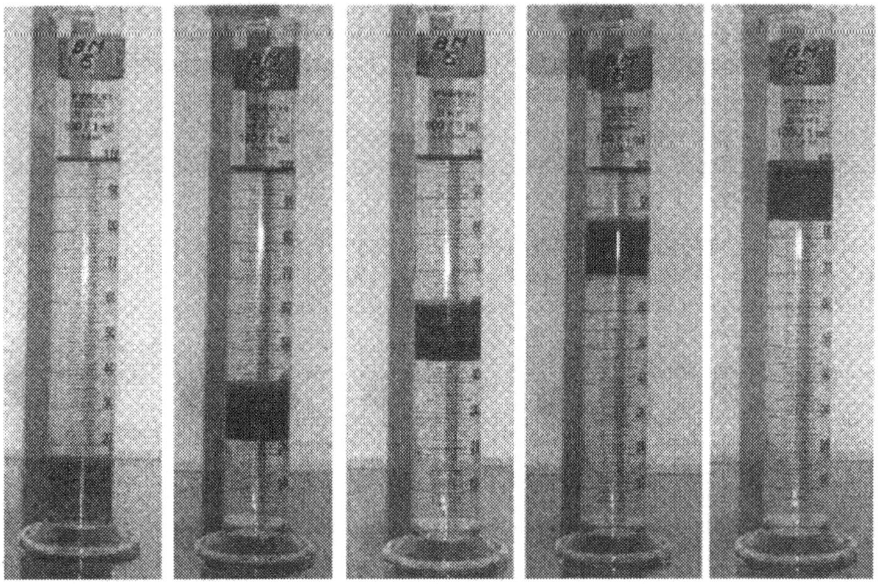

Figure 2: Floating of the granules after settling. Fig2A to Fig.2 E correspond to 2, 24, 25, 28, and 32 min after settling of the granules. The sample was collected at the end of one cycle of cultivation in SBR fed with synthetic wastewater which was composed of 1000 mg l-1 of COD (ethanol), 300 mg l-1 of ammonia nitrogen, 2400 mg l-1 of bicarbonate, and micronutrients (Yang et al., 2003). Almost all ammonia was oxidized to nitrate by the end of cycle.

Formation of granules in SBR

The granules were retained in the SBR while the flocs were washed out with the effluent. Concentration of granular biomass (VSS) during 6 days of experimental period was stable, at 6.5 ± 0.2 g L-1. Concentration of floc biomass (VSS) was 0.15 ± 0.02 g L-1. Stable concentration of granular biomass can be due to balanced attachment and detachment of the flocs to granules or balanced growth and destruction of the granules. The hydraulic residence time was 0.33 d, which corresponded to a daily exchange of 3 reactor volumes. Therefore, the ratio of produced granular biomass to produced floccular biomass was 14.5. This ratio was close to 18.3, the initial ratio of granular labeled biomass to the flocculent labeled biomass after 4 h of labeling with lipophilic tracer (one growth cycle in SBR). Content of lipophilic tracer in granular biomass was stable for 6 days of study and was thought to be because of balanced attachment and detachment of the flocs to granules or balanced growth and destruction of the granules. It cannot be the result of negligible degradation of granules because the labeled biomass permanently released as the labeled flocs. The tracer content could decrease if the rate of granule growth is higher than the rate of granule degradation.

Duration of the time of granule formation from activated sludge is usually from 8 to 14 days (Fig.3). This duration can be diminished if selected microbial cultures with high self-aggregation ability are added to the SBR. The strains of aerobic bacteria with aggregation index (AI) higher than 60-80 % or with cell hydrophobicity, measured by hydrocarbon adherence test, higher than 80%, were selected and isolated from the microbial granules using the repeated cycles of adhesion, settling, and cultivation. There were sporogenic Gram-positive rods, Gram-negative rods, and gliding bacteria. The granules with mean diameter 1 mm were formed after 2 days when cells with high cell surface hydrophobicity were used as inoculum (seeds) for granulation (Fig.4). Formation of the granules with mean diameter 1 mm from the flocs of activated sludge required 8 days.

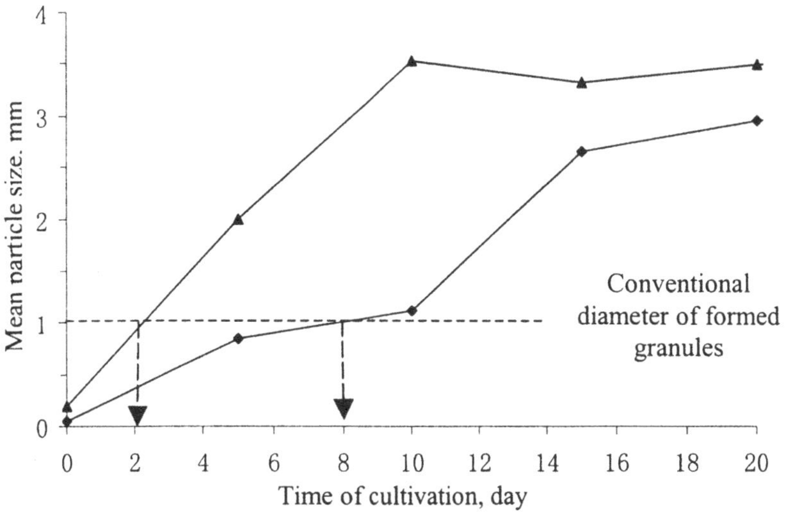

Figure 3: Formation of the granules from selected cells with high cell surface hydrophobicity (1) and from the flocs of activated sludge

Conclusions

1. Granules typically consisted of an external layer of obligate aerobic microorganisms, followed by facultative anaerobic, then obligate anaerobic bacteria (at a depth of 800 mm below the granule surface), and finally a core of dead and lysed cells.
2. The optimal diameter of the studied aerobic granule is less than 1.7 mm considering absence of the layer of obligate anaerobic bacteria or less than 0.5 mm considering that the whole granule should have a porous biomass-filled matrix.
3. Time of granule formation from the flocs of activated sludge was from 8 to 14 days but can be diminished to 2 days if selected bacterial strains with enhanced self-aggregation will be used instead of activated sludge.

Acknowledgements

This research was supported by Nanyang Technological University, Singapore.

References

Amann R.I., Krumholz L. and Stahl, D.A. (1990). Fluorescent-oligonucleotide probing of whole cells for determinative, phylogenetic, and environmental studies in microbiology. J. Bacteriol., 172 (2), 762-770.

Beun J.J., Hendriks A., van Loosdrecht M.C.M., Morgenroth E., Wilderer P.A. and Heijnen J.J. (1999). Aerobic granulation in a sequencing batch reactor. Wat. Res., 33 (10), 2283-2290.

Beun J.J., van Loosdrecht M.C.M. and Heijnen J.J. (2000). Aerobic granulation. Wat. Sci. Technol., 41 (4-5), 41-48.

Beun J.J., van Loosdrecht M.C.M. and Heijnen J.J. (2002). Aerobic granulation in a sequencing batch airlift reactor. Wat.Res., 36 (3), 702-712.

Etchebehere C., Errazquin M. I., Cabezas A., Pianzzola M.J., Mallo M., Lombardi P., Ottonello G., Borzacconi L., and Muxí L. (2002). Sludge bed development in denitrifying reactors using different inocula-performance and microbiological aspects. Wat. Sci. Technol., 45 (10), 365-370.

Etterer T. and Wilderer P.A. (2001). Generation and properties of aerobic granular sludge. Wat. Sci. Technol., 43 (3), 19-26.

Gieseke A., Purkhold U., Wagner M., Amann R. and Schramm A. (2001). Community structure and activity dynamics of nitrifying bacteria in a phosphate-removing biofilm. Appl. Environ. Microbiol., 67 (3), 1351–1362.

Massol-Deya A.A., Whallon J., Hickey R.F. and Tiedje, J.M. (1995). Channel structures in aerobic biofilms of fixed-film reactors treating contaminated groundwater. Appl. Environ. Microbiol., 61 (2), 769-777.

Morgenroth E., Sherden T., van Loosdrecht M.C.M., Heijnen J.J and Wilderer P.A. (1997). Aerobic granule sludge in a sequencing batch reactor. Water Res., 31 (12), 3191-3194.

Moy B.Y.P., Tay J.H., Toh S.K., Liu Y. and Tay S.T.L. (2002). High organic loading influences the physical characteristics of aerobic sludge granules. Lett. Appl. Microbiol., 34, 407-412.

Okabe S., Satoh H., Watanabe Y. (1999). In situ analysis of nitrifying biofilms as determined by in situ hybridization and the use of microelectrodes. Appl. Environ. Microbiol., 65 (7), 3182-3191.

Peng D.C., Bernet N., Delgenes J.-P. and Moletta R. (1999). Aerobic granular sludge - a case report. Wat. Res., 33 (3), 890-893.

Sghir A., Gramet G., Suau A., Rochet V., Pochart P. and Dore J. (2000). Quantification of bacterial groups within human fecal flora by oligonucleotide probe hybridization. Appl. Environ. Microbiol., 66 (5), 2263- 2266.

Tay J.H., Liu Q.S. and Liu Y. (2001a). Microscopic observation of aerobic granulation in sequential aerobic sludge reactor. J. Appl. Microbiol., 91 (1), 168-175.

Tay J.-H., Liu Q.-S. and Liu Y. (2001b). The role of cellular polysaccharides in the formation and stability of aerobic granules. Lett. Appl. Microbiol., 33 (3), 222-227.

Tay J.-H., Ivanov V., Pan S. and Tay S. T.-L. (2002a). Specific layers in aerobically grown microbial granules. Lett. Appl.Microbiol., 34 (4), 254-258.

Tay S.T-L., Ivanov V., Yi S., Zhuang W.-Q. and Tay J.-H. (2002b). Presence of anaerobic Bacteroides in aerobically grown microbial granules. Microbial Ecology, 44 (3), 278-285.

Tay J.-H., Tay S. T.-L., Ivanov V., Pan S., Jiang H.-L. and Liu Q.-S (2003a). Distribution of biomass and porosity profiles in microbial granules used for aerobic wastewater treatment. Lett. Appl. Microbiol., 36 (5), 297-301.

Tay J.-H., Pan S., Tay S.T.L., Ivanov V. and Liu Y. (2003b). The effect of organic loading rate on the aerobic granulation: the development of shear force theory. Wat.Sci. Technol., 47 (11), 235–240.

Toh S.K., Tay J.H., Moy B.Y.P., Ivanov V. and Tay S.T.L. (2002). Size effect on the physical characteristics of the aerobic granule in a SBR. Appl. Microbiol. Biotechnol., 60 (6) , 687-695.

Villaverde S. and Fernandez-Polanco F. (1999). Spatial distribution of respiratory activity in Pseudomonas putida 54G biofilms degrading volatile organic compounds (VOC). Appl. Microbiol. Biotechnol., 51(3), 382-387.

Yang S.-F., Tay J.-H. and Liu Y. (2003). A novel granular sludge sequencing batch reactor for removal of organic and nitrogen from wastewater. J. Biotechnol., 106 (1), 77-86.
Zhu J.-R. and Wilderer P. A. (2003). Effect of extended idle conditions on structure and activity of granular activated sludge. Wat. Res., 37 (9), 2013-2018.

Population Dynamics during Aerobic Granule Formation: Lessons from denaturing gradient gel electrophoresis

B. S. McSwain[*,**]**, R. L. Irvine**[*]**, P. A. Wilderer**[**]

* Department of Civil Engineering and Geological Sciences, University of Notre Dame, Notre Dame, IN 46556, USA (E-mail: bmcswain@nd.edu; Irvine.1@nd.edu)
** Institute of Water Quality Control and Waste Management, Technical University of Munich, 85748 Garching, Germany (E-mail: P.Wilderer@bv.tum.de)

Abstract Many factors influence the formation and structure of aerobic granular sludge in the Sequencing Batch Reactor (SBR). Short or static FILL periods create a high feast condition, which selects for organisms that have high substrate uptake kinetics, such as floc-formers. Alternatively, short setting times are used to continually wash out slow settling flocs, so only aggregate forming organisms remain in the reactor. Hitherto, the influence of feast-famine and settling time on species selection was discussed in hypothetical terms. In this experiment, four identical SBRs were operated with two settling times (2 and 10 min) and three types of FILL periods (0, 33, and 66% Aerated FILL). By altering the ratio of static and aerated FILL, the feast-famine condition was varied. The population dynamics and stability were determined using denaturing gradient gel electrophoresis (DGGE) and image analysis. As the ratio of aerated FILL increased, filamentous content in granules increased, and the population diverged faster from the inoculum. The steady-state community of all three granule reactors contained only a few dominant species, with some species being common, showing that selection was not random. Only the reactor with completely static FILL and a short settling time formed smooth granules, and this community was the most stable over sixty days of steady-state operation (96% similarity). The reactor with a longer settling time formed a mixture of flocculent and granular sludge, and the population was much more diverse and less stable than the granular reactor.

Keywords: Sequencing Batch Reactor, SBR, Aerobic Granules, DGGE, Population Dynamics

Introduction

The ultimate success of wastewater treatment depends, first, upon the selection of metabolically capable microorganisms and, second, upon the efficient separation of those organisms from the treated effluent. Much research has focused on reducing the settling time required for activated sludge by forming dense flocs or by using biofilm reactors. Recently, aerobic granular sludge has been presented as a fast settling bacterial aggregate that can be defined as a self-immobilized biofilm. Aerobic granules have been formed in several Sequencing Batch Reactors (SBRs) treating synthetic, municipal, and industrial wastewaters (Beun et al., 2002; Morgenroth et al., 1997; Schwarzenbeck et al., 2004).

The cause and mechanism of granulation are not yet fully understood, but several important factors have been described in literature. In order to form spherical, compact granules, the conversion of biodegradable substrate into inter-cellular stored substrate must occur rapidly in the reactor. This is often described as a feast-famine regime, and can easily be applied in the SBR with short or static fill periods (McSwain *et al.*, 2004a; de Kreuk *et al.*, 2004). Generally, floc-forming bacteria, with relatively high substrate uptake kinetics, have an advantage over filamentous bacteria if food is supplied intermittently, forcing bacteria to acquire and store the food for cell maintenance and growth during periods of starvation (Chudoba *et* al., 1973). By creating a feast-famine environment, the SBR influences population dynamics and the selection of aggregate forming organisms (Chiesa *et al.*, 1985).

Additional operating parameters affect granule formation, including the dissolved oxygen concentration (de Kreuk *et al.*, 2004), shear force (Tay *et al.*, 2001a), and settling time (Beun *et al.*, 1999; 2002). Short settling times have been utilized to select for fast settling flocs and granules, forcing the washout of less dense flocs and suspended organisms (Beun *et al.*, 1999; 2002). During reactor start-up, the feast-famine regime and short settling time are thought to greatly influence the selection of organisms in the reactor, which ultimately controls the sludge properties and reactor performance.

In order to optimize granule formation, engineers must first understand the microbial composition, structure, and stability of the microbial population. In the past, investigations of the microbial community required microbiological techniques and conventional microscopy that were time intensive and had severe limitations. Microscopy of natural samples can be complicated by soil and sediment particles, and general information about species' identity requires a top to bottom approach using fluorescent gene probes, as applied with FISH (fluorescent in situ hybridization). Researchers could attempt selective enrichment through culturing, but not all organisms grow on media in the laboratory. Therefore, scientists' understanding of bacterial diversity and stability in a mixed, natural sample has been limited by the available laboratory techniques. Denaturing gradient gel electrophoresis (DGGE) is a recent advancement that allows for the fast, direct visualization of the microbial diversity in an environmental sample. By obtaining genetic fingerprints of samples over time, the population dynamics and stability of a community can be followed (Muyzer *et al.*, 1998).

This paper reviews the population dynamics from two granular reactor experiments. In the first experiment, the SBR Fill condition was varied to determine the effect of feast-famine on species selection and the formation of granules. In the second experiment, two different settling times were applied to select for one granular and one flocculent reactor. The sludge formed in each reactor was characterized using simple microscopy, and the reactor performance was described by typical engineering parameters. Species selection and stability was studied using DGGE, and the results are presented herein.

Methods

Reactors were fed from a common feed of glucose, peptone, and nutrients (McSwain *et al.*, 2004a). All reactors received 800 mg COD L^{-1} every cycle or a loading rate of 2.4 kg COD m^{-3} day^{-1}. The SBRs were operated on a four-hour cycle with variations in either the ratio of Static to Aerated FILL (Reactors 1, 2, and 3) or with different settling times (Reactors 1_S10 and 1). The original inoculum was collected from the activated sludge basin of a municipal wastewater treatment plant (Garching, Germany) and was aerated for twenty-four hours in batch before being divided into lab-scale SBRs.

Mixed liquor and volatile suspended solids (MLSS and VSS) content, effluent solids and volatile solids (ESS and EVSS) content, and the sludge volume index (SVI) were measured according to APHA standard engineering methods (*Standard Methods*, 1998). Substrate removal was measured by the chemical oxygen demand (COD) over a cycle and in the effluent using Dr.Lange COD kits (following the colorimetric COD standard method). The development of flocs and granules were observed using a stereomicroscope (Leica Wild MPS 46/52), and images were obtained with an attached Kodak digital camera.

Table 1: Operating Strategy (Variation of fill phase to alter feast-famine condition)

Cycle Phase	**Cycle Times (min)**			
	R1_S10	**R1**	**R2**	**R3**
Static Fill (Fill, no aeration, no mixing)	90	90	60	30
Aerated Fill (Fill, aeration, mixing)	0	0	30	60
React	120	120	120	120
Settle	10	2	2	2
Draw	15	15	15	15
Idle	5	13	13	13
Total Cycle Time	240	240	240	240

For community analysis, reactor samples were homogenized, washed twice in 0.85% KCl and extracted using a Bio 101 DNA for Soil extraction kit. 50 ng DNA was amplified using the 318F-GC clamp and 541R 16S rRNA primers as described in Muyzer *et al.* (1998) and Qiagen HotStart Taq Polymerase with recommended reagent mixes. PCR proceeded with 15 min activation at 94°C, 30 cycles each of 94°C denaturation (1 min), 55°C annealing (1 min), 72°C elongation (1 min), and a final step of 72°C for 10 min. 20 μL of product was loaded with 10 μL loading dye to a 8% acrylamide, 40-60% denaturing gel and run for 16 hours at 100V. DGGE gels were stained with Sybr Gold (Molecular Probes, Oregon, USA), and cluster analysis of gel lanes were performed using GelCompar II software (Applied Maths, Belgium).

Individual bands of interest were cut from the stained DGGE gal, and DNA was extracted using the Qiaex II Kit (Qiagen, Hilden, Germany). Extracted DNA was re-amplified and separated using PCR-DGGE as described above. The same band of interest was then cut again, DNA extracted from the gel, and re-amplified using a 341F / 518R primer pair (no GC clamp). The PCR products were cleaned using the NucleoSpin Extract Kit (Macherey-Nagel, Germany), and sequencing ws performed by MWG (Germany) with 341F (no GC clamp) used as a sequencing primer. Sequences were matched using the NCBI nucleotide-nucleotide BLAST search (http://www.ncbi.nlm.nih.gov/blast/). The phylogenetic hierarchy of the best sequence to sequence known organism match was determined using the hierarchy browser of the Ribosomal Database Project (http://rdp.cme.msu.edu/index.jsp).

Results & Discussion

The Effect of Feast-Famine

An in depth discussion of the performance of Reactors 1, 2, and 3 is provided in McSwain *et al.* (2004a). Each SBR yielded predominantly granular sludge within one month of operation, and all reactors had stable and complete COD removal (> 96%). Figure 1 shows the COD accumulation and removal over one cycle on day 30 of operation. The theoretical COD line represents the concentration of COD fed to all reactors over 90 minutes if no microbial consumption takes place. However, since the aeration began after 30 minutes for Reactor 3 and after 60 minutes for Reactor 2, the plot for each reactor shows COD removal beginning at these times and continuing during the remaining Aerated FILL. Figure 1 also shows how the differential between feast and famine was varied in the three reactors by altering the ratio of Static to Aerated FILL. Reactor 1 accumulated the total COD before aeration began, providing the highest feast condition.

The granules grew and changed considerably in size and morphology over several months, and pseudo-steady state was established after four months based upon MLSS and SVI measurements. The MLSS content decreased steadily for Reactors 1 through 3 (9.0, 6.8, and 3.2 g L^{-1}, respectively). The SVIs for Reactors 1, 2 and 3 increased from 46 mL g^{-1} SS for Reactor 1 to 60 and 114 mL g^{-1} SS, for Reactors 2 and 3. The effluent solids content also increased for Reactors 1 through 3, ranging from 170 mg L^{-1} ESS to 220 and 290 mg L^{-1} ESS for each reactor.

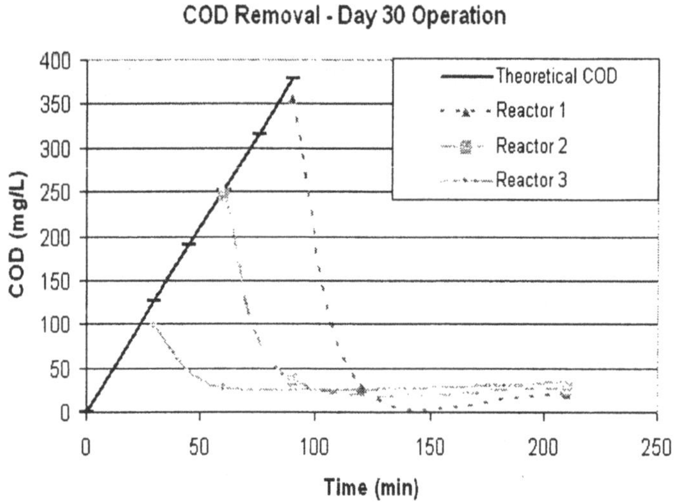

Figure 1: COD removal during one cycle for reactors 1, 2, and 3. Variations in Static and Aerated FILL create different feast-famine environments. Reactor 1 with 100% Static Fill has the highest COD concentration, while Reactor 3 has the lowest.

After 60 days of operation, all reactors contained predominantly granule sludge. The structure of the granules was clearly different in terms of filamentous bacteria and settling ability. Microscopic observations confirmed that the filamentous content of granules increased with increasing times for Aerated Fill (or, as is shown in Fig.1, a smaller feast-famine differential). Typical granules from each reactor at pseudo steady-state are shown in Figure 2.

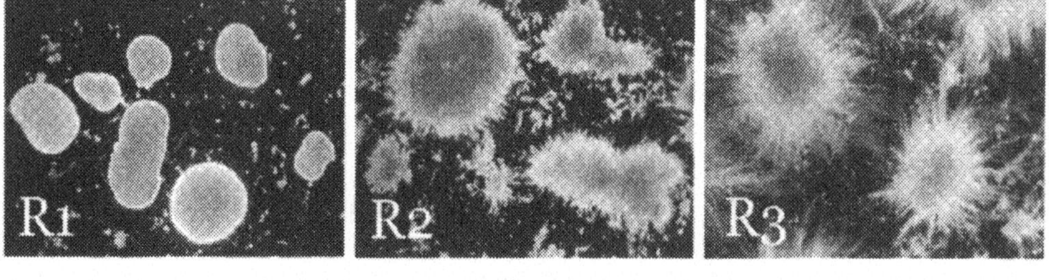

Figure 2: Images of aerobic granules cultivated in each reactor (Reactor 1, 2 and 3 with 0%, 33% and 66% Aerated Fill, respectively). Scale bar = 1mm.

It is clear from microscopy, MLSS content, and SVI values that granule structure is affected by intermittent or extended feeding. Dense, smooth granules were only formed in Reactor 1, which simulated a dump fill and had the greatest differential between feast and famine. With extended aerated feeding periods in Reactors 2 and 3, filamentous organisms were evident in granule structure. These observations correlate well with previous research by Chudoba *et al.* (1973) and Chiesa *et al.* (1985) in suspended growth systems. This research showed that floc-forming bacteria had a selective advantage over filamentous organisms due to substrate-uptake kinetics and starvation sensitivity. Filamentous organisms were able to compete for substrate in Reactors 2 and 3 since feeding continued during aeration, whereas only the organisms with the highest substrate uptake kinetics competed in Reactor 1 with a stringent intermittent feeding program. The current experiment demonstrates that principles of species selection apply to granule reactors or self-immobilized biofilm systems and have a significant impact on granular structure.

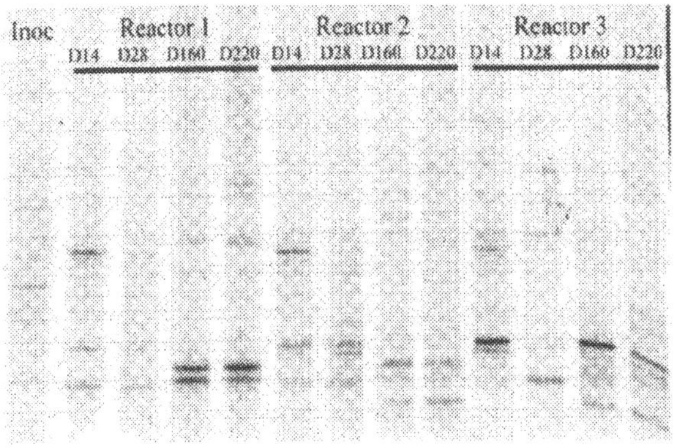

Figure 3: DGGE gel of samples taken from Reactors 1, 2, and 3 over time.

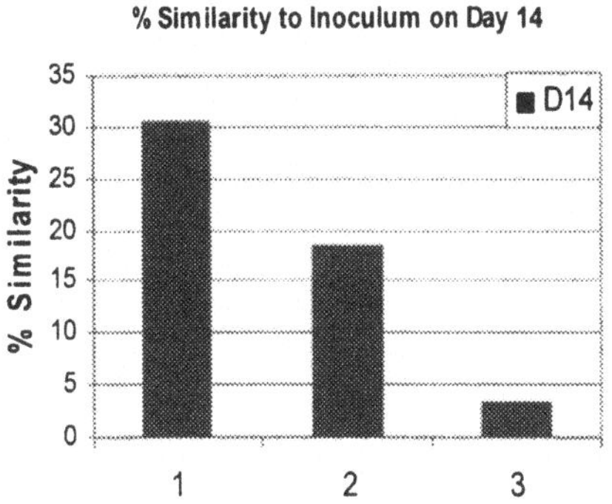

Figure 4: % Similarity of the Day 14 microbial community to the original inoculum.

Table 2: Similarity of steady-state samples.

	R1-160	R1-220	R2-160	R2-220	R3-160	R3-220
R1-160	100					
R1-220	96	100				
R2-160	71	76	100			
R2-220	48	52	83	100		
R3-160	14	15	20	24	100	
R3-220	24	27	18	14	75	100

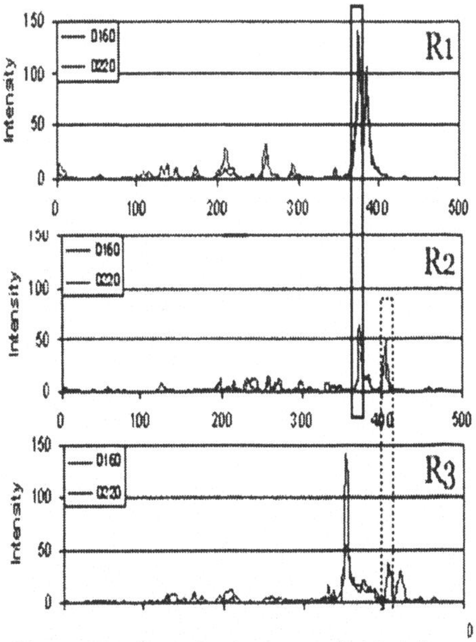

Figure 5: Densitograms of steady-state samples. Boxes show common organisms in R1 & R2, and R2 & R3.

To visualize selection in these reactors during reactor start-up, the bacterial genetic fingerprints from DGGE gels were compared using image analysis software. One composite DGGE gel is shown in Fig.3. To access the rate of population divergence from the inoculum during granule formation, the similarity coefficient was calculated for the first two and four week samples (days 14 and 28). Figure 4 shows that with an increased percentage of Aerated FILL, the microbial community in Reactors 1 through 3 showed decreasing similarities to the inoculum after two weeks. With increased Aerated FILL, the microbial population diverged faster from the inoculum, indicating that extended aeration during FILL drove the establishment of a different microbial community than what was present in the inoculum. However, the rate of microbial selection in a mixed sample will depend upon the starting microbial community and the selection pressures applied to the system.

As a measure of population stability at steady-state, two samples from days 160 and 220 were compared. The similarity matrix is shown in Table 2. Overall, the granular Reactor 1 had the greatest steady-state stability, 96%, and the stability decreased with increasing Aerated FILL (83% and 75% for Reactors 2 and 3, respectively). It is notable that only two strains were dominant in each reactor at steady-state, and Reactor 2 shared one of these strains with Reactor 1 and 3 (shown in Fig.5). Reactor 2, with an extended Aerated FILL and a medium feast condition, provided selection pressures for both flocs and filaments. However, Reactor 2 was more similar to Reactor 1 than Reactor 3 (see Table 2).

Dominant bands in the DGGE profiles of each reactor were excised, purified, and sequenced, and their identification and phylogenetic relationships are presented below in Table 3. Based on DGGE gel positioning, Reactors 1 and 2 had bands R1-1 and R2-5 in common, and Reactors 2 and 3 had bands R2-6 and R3-10 in common. Sequencing showed that these organisms corresponded to *Ferribacterium limneticum* (Reactors 1 and 2) and *Thiothrix* (Reactors 2 and 3). *Ferribacterium* is in the Rhodocyclaceae family, of which the classic floc-former *Zoogloea* also belongs. This was one of the dominant species in both R1 and R2 but not in the filamentous R3 reactor. Both dominant species in R3 were filamentous organisms. *Thiothrix* is a common filament in wastewater treatment plants, and it was selected in both Reactors 2 and 3 The second dominant band in R3 corresponded to *Leptothrix mobilis*, which is in the same genus as the well-known filament *Sphaerotilus*.

Table 3: Phylogenetic relationship of excised DGGE bands.

Band	Highest Similarity	Phylogenetic Family Group	Identity Match
R1-1	*Ferribacterium limneticum*	Rhodocyclaceae	120/141
R1-2	*Hydrogenophaga taeniospiralis*	Comamonadaceae	141/154
R1-3	*Bacteroidetes* bacterium KMM 3914	Flavobacteriaceae	77/87
R2-4	Uncultured bacterium MK12	Bacteroidetes*	122/136
R2-5	*Ferribacterium limneticum*	Rhodocyclaceae	125/143
R2-6	*Thiothrix* sp. OS-F	Thiotrichaceae	92/95
R3-7	*Micrococcus luteus* strain SAFR-002	Micrococcineae	122/129
R3-8	*L.mobilis*	Incertae sedis	118/138
R3-10	*Thiothrix* sp. CT3	Thiotrichaceae	97/100

*only phylum level identification available

By using DGGE in this experiment, it is clear that Aerated FILL affects granule formation through species selection. Moreover, this selection is not random, since Reactors 1 and 2 and 2 and 3 selected for common organisms. While all steady-state samples showed a complex fingerprint, there were generally two dominant species in each reactor. Finally, at steady-state, smooth granules in Reactor 1 have the most stable population.

The Effect of Settling Time

A detailed description of the reactor performance and operation is presented in McSwain *et al.* (2004b). Due to the difference in settling time, the washout of sludge during the first two weeks of operation was much greater for Reactor 1 (2 min settling) than for Reactor 1_S10 (10 min settling). After 80 days of operation, the reactors began to diverge in terms of sludge characteristics, MLSS, and SVI. For Reactor 1_S10, any observed granules always co-existed with flocculent sludge during one hundred days of steady-state operation. Figure 6 shows the steady-state sludge for Reactor 1_S10, while sludge from Reactor 1 was presented in Fig.2. Both reactors performed well in terms of COD removal and oxygen uptake rate. The reactors differed in the properties of the sludge, MLSS content, and settling characteristics. Most significantly, the flocculent Reactor 1_S10 had an MLSS of 3.0 g L^{-1} and SVI of 115 mL g^{-1}, compared to an average MLSS of 9.0 g L^{-1} and an SVI of 46 mL g^{-1} in the granular Reactor 1.

Figure 6: Steady-state sludge from Reactor 1_S10. Scale bar = 1 mm.

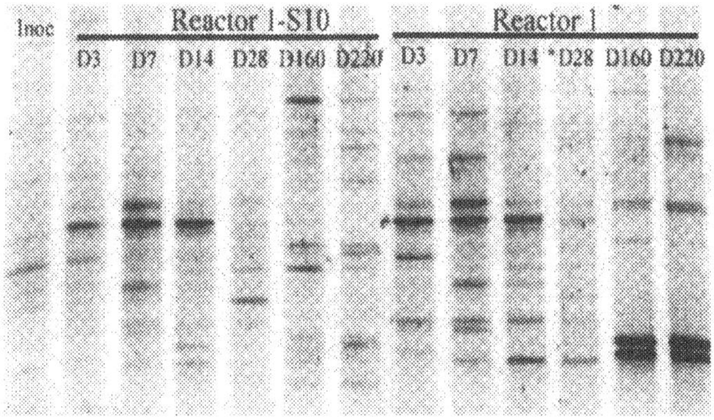

Figure 7: DGGE gel showing samples from R1_S10 and R1 taken over time.

To address the affect of washout on species selection, DGGE genetic fingerprints from of the reactor communities over time were compared (shown in Fig.7). The first four lanes for each reactor show the population dynamics during the first month of operation. By Day 28, it is clear that some dominant bands, that were common to both reactors, lose their intensity. Likewise, differences arise between the dominant bands in Reactor 1_S10 and 1). The last two lanes for each reactor show samples taken from steady-state operation, 60 days apart. These lanes represent the stability and diversity of steady-state sludge. The normalized densitograms were produced using image analysis, and these are presented in Fig.8. The gray line represents Day 160, and the black line represents Day 220. The similarity of the 4 lanes was calculated, and the matrix is presented in Table 4. Steady-state samples from the flocculent Reactor 1_S10 were more diverse than the granular Reactor 1, and the community showed less long term stability (only 57% for Reactor 1_S10 steady-state). Reactor 1 selected for 4 main bacterial species, and the genetic steady-state fingerprints were 96% similar. Overall, the communities from R1_S10 and R1 were only 19 and 50% similar on days 160 and 220, respectively. This clearly shows that settling time causes differences in species selection and community stability.

Table 4: % Similarity based on Pearson correlation of DGGE fingerprints from steady-state samples.

	R1_S10 D160	R1_S10 D220	R1 D160	R1 D220
R1_S10 D160	100			
R1_S10 D220	57	100		
R1 D160	19	49	100	
R1 D220	18	50	96	100

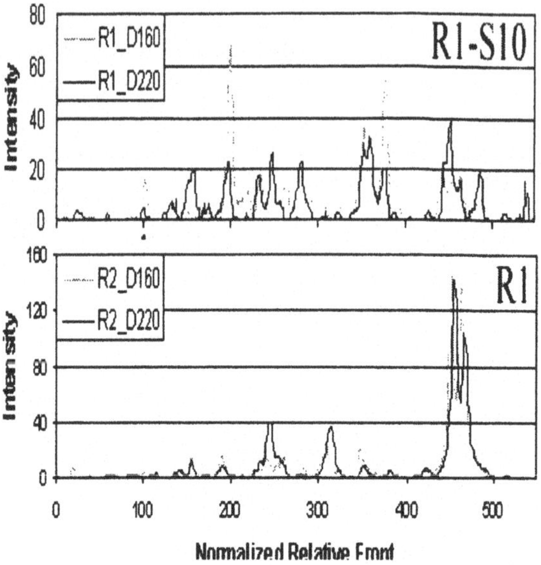

Figure 8: Normalized densitograms of steady-state DGGE patterns.

Conclusions

As the ratio of aerated FILL increases, filamentous content in granules increases. Therefore, a high feast-famine condition is necessary for smooth, compact granule formation. Extended feeding during aeration creates a strong selection pressure for filamentous bacteria. This was shown by the increasing rate of population divergence from the inoculum for Reactors 2 and 3. The steady-state community of all three granule reactors contained only a few dominant species, with some species being common. This indicates that selection was not random.

Only the reactor with completely static FILL and a short settling time formed smooth granules, and this community was the most stable over sixty days of steady-state operation (96% similarity). The reactor with a longer settling time formed a mixture of flocculent and granular sludge, and the population was much more diverse and less stable (57%) than the granular reactor at steady-state.

Acknowledgements

This research was funded by the US Department of Education and the German Research Foundation. The authors thank the Institut für Bodenökologie, GSF Forschungszentrum für Umwelt und Gesundheit, Neuherberg, for the use of the Gelcompar II software.

References

Beun J.J., Hendriks A., van Loosdrecht M.C.M, Morgenroth E., Wilderer P.A. and Heijnen J.J. (1999). Aerobic granulation in a sequencing batch reactor. *Wat. Res.*, **33** (10), 2283-2290.

Beun J.J., van Loosdrecht M.C.M and Heijnen J.J. (2002). Aerobic granulation in a sequencing batch airlift reactor. *Wat. Res.*, **36**, 702-712.

Chiesa S.C. and Irvine R.L. (1985). Growth and control of filamentous microbes in activated sludge: an integrated hypothesis. *Wat. Res.*, **19**(4), 471-479.

Chiesa S.C., Irvine R.L. and Manning J.F. Jr. (1985). Feast/famine growth environments and activated sludge population selection. *Biotechnol. and Bioeng.*, **27**, 562-569.

Chudoba J., Grau P. and Ottová V. (1973). Control of activated sludge filamentous bulking – II. Selection of microorganisms by means of a selector. *Wat. Res.*, **7**, 1389-1406.

de Kreuk M.K. and van Loosdrecht M.C.M. (2004). Selection of slow growing organisms as a means for improving aerobic granular sludge stability. *Wat. Sci. Tech.*, 49 (11-12), 9-17.

McSwain B.S., Irvine R.L. and Wilderer P.A. (2004). The effect of intermittent feeding on aerobic granule structure. *Wat.Sci.Tech.*, 49 (11-12), 19-25.

McSwain B.S., Irvine R.L. and Wilderer P.A. (2004). The influence of settling time on the formation of aerobic granules. IWA SBR 3 Conference, Noosa, Australia.

Morgenroth E, Sherden T, van Loosdrecht MCM, Heijnen JJ and Wilderer PA. (1997).Aerobic granular sludge in a sequencing batch reactor. *Wat. Res.*, **31,** 3191–3194.

Muyzer G., Brinkhoff T., Nubel U., Santegoeds C., Schafer H., and Wawer C (1998). Denaturing gradient gel electrophoresis (DGGE) in microbial ecology. *Mol. Microb. Ecol. Manual*, **3.4.4**, 1-27.

Standard Methods for the Examination of Water and Wastewater (1998). 20th ed., American Public Health Association/American Water Works Association/Water Environment Federation, Washington DC, USA.

Schwarzenbeck N., Erley R. and Wilderer P.A. (2004). Aerobic granular sludge in an SBR-system treating wastewater rich in particulate matter. *Wat. Sci. Tech.*, **49** (11-12), 41-46.

Tay J.-H., Liu Q.-S. and Liu Y. (2001). The effects of shear force on the formation, structure and metabolism of aerobic granules. *Appl. Microbiol. Biotech.*, **57**, 227-233.

Tay J.-H., Liu Q.-S. and Liu Y. (2001). The role of cellular polysaccharides in the formation and stability of aerobic granules. *Letters Appl. Microbiol.*, **33**, 222-226.

Nitrifying granular sludge in a Sequencing Batch Reactor

Mosquera-Corral A.*, Vázquez-Padín J.R., Arrojo B., Campos J.L. and Méndez R.

* Department of Chemical Engineering. Institute of Technology. University of Santiago de Compostela, E-15782, Santiago de Compostela, Spain. E-mail address: eqanusmc@usc.es

Abstract An aerobic granular sequencing batch reactor was operated at different COD/N ratios to study its nitrogen removal efficiency. Maximum removal percentages of 90% for the organic matter and up to 55% for the nitrogen were achieved. Removal of ammonia was carried out by both assimilation and simultaneous nitrification-denitrification processes. The predominance of each mechanism was related to the COD/N ratio in the feeding media. In spite of the changes in the feeding composition the granules maintained their structures and the solids content in the effluent was reduced to 10 mg TSS/L when acetate was removed from the feeding media. Stable nitrifying granular sludge was obtained from heterotrophic granular sludge by decreasing the COD/N ratio.

Keywords COD/N ratio, Denitrification, Granules, Nitrification, Nitrogen removal

Introduction

Activated sludge is one of the most common biological processes used for wastewater treatment of organic and nitrogenous compounds. Usually, in these units, the loading rate is limited by the maximal biomass concentration that can be maintained inside the system. To maintain high biomass concentrations inside the system settlers with very large areas are required. Besides the nitrification process is the limiting step for the treatment of nitrogenous compounds due to the slow growing rate of the microorganisms involved. This fact provokes the necessity of a high solid retention time (SRT) and, therefore, a large volume of the system.

To improve the sludge retention time (SRT) it is possible the immobilization of microorganisms in form of biofilms or granules. Higher biomass concentrations and lower unit volumes are achieved with immobilization. A biofilm airlift suspension (BAS) reactor is an example of this kind of reactors. The processes of COD removal and both nitrification and denitrification can occur simultaneously in a continuously fed BAS reactor (Tijhuis *et al.*, 1994; Van Benthum *et al.*, 1997). Recently aerobic granular sludge was developed in a sequencing batch airlift reactor (SBAR) (Beun *et al.*, 1999) where the removal of COD as well as nitrification and denitrification processes occurred (Beun *et al.*, 2001).

In general, the population distribution in biofilms is the result of the difference in growth rates. The slow-growing nitrifiers are located inside the biofilms, the fast-growing heterotrophs are located more in the outer layer of the biofilms (van Loosdrecht *et al.*, 1995). In a discontinuously fed system, the organic carbon source penetrates completely into the biofilms because of the temporarily high concentration in the liquid. Dissolved oxygen is present only in the outer layer of the biofilms due to its consumption rate. Consequently, it can be expected that the nitrifiers are located in the outer aerobic layer of the biofilms (de

Beer *et al.*, 1993) together with the heterotrophs oxidizing aerobically the organic matter, and the organic carbon compounds are stored anoxically as PHB (Poly hydroxy butirate) by the heterotrophs inside the biofilms. In the inner core of the granule the heterotrophs grow using organic matter and reducing nitrate generated by nitrification to nitrogen gas. They were also present in the outer layers of the granules where they could use oxygen as electron acceptor.

In the present work an aerobic granular sequential batch reactor was operated using different COD/N ratios in the feed to obtain a mainly nitrifying reactor. Effluents with different compositions in nitrogen compounds were generated. The effects of the COD/N ratio of the fed on the overall nitrogen removal efficiency were tested.

Methods

Experimental set-up

A sequencing batch reactor (SBR) with a total volume of 2.5 L and a working volume of 1.5 L was used. Dimensions of the unit were: height of 465 mm and inner diameter of 85 mm, the height to the diameter ratio (H/D) being 5.5. The maximum level of the liquid was 264 mm, and the minimum level 132 mm after effluent withdrawal. Oxygen was supplied by using spargers to promote the formation of small air bubbles. A set of two peristaltic pumps was used to feed and to discharge the effluent, respectively. The influent was introduced in the system through ports located at the top of the reactors. The effluent was discharged through the sampling port placed at middle height of the column reactor (Fig. 1). A programmable logic controller (PLC) Siemens model S7-224CPU controlled the actuations over the pumps and valves, and thus the length of every operational period in the SBR. The reactor was operated at room temperature (15-20 °C), at oxygen concentrations of 6-8 mg O_2/L (during the feast period) and 8 mg O_2/L (during the famine period) and without pH control, which varied between 7.4 and 8.5.

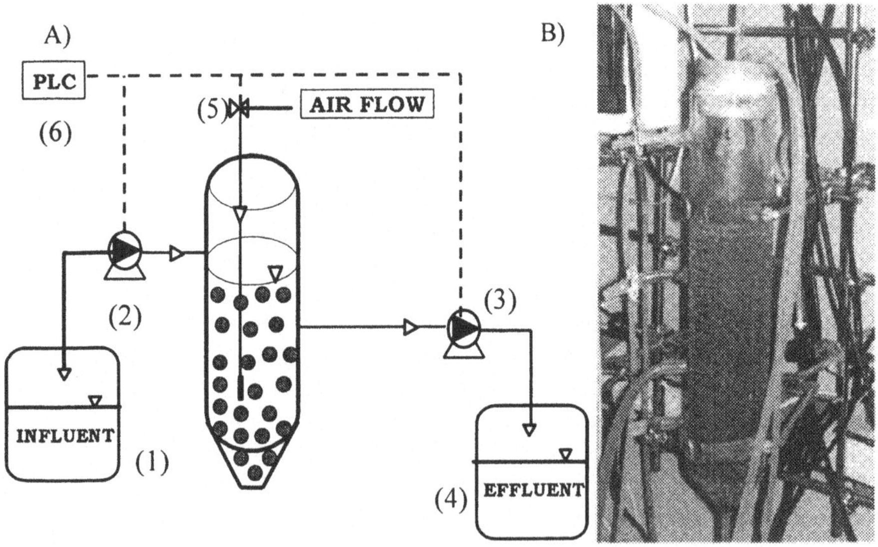

Figure 1: A) Experimental set up: (1) Feeding tank; (2) Feeding pump; (3) Effluent pump; (4) Effluent tank (5); Air valve; (6) PLC. B) Picture of the reactor.

Strategy of operation

The SBR was fed with a synthetic solution with the composition described in Table 1 according to Beun *et al.* (1999) and the trace solution to Smolders *et al.* (1995). The system was inoculated with a granular sludge collected from another SBR.

Table 1: Composition of the synthetic wastewater used to feed the SBR.

Parameters	Concentration (mg/L)
COD	0-1000
NH_4^+-N	30-200
PO_4^{3-}-P	23
Traces	1 mL/L

The operational conditions of the reactor are summarized in Table 2. Different COD/N ratios of 15.0, 7.0, 5.0, 2.5, 1.25 and 0 g/g in the feeding were tested.

The system was operated in cycles of 3 hours with an exchange volume of 50%. Every cycle comprehended a feeding period without aeration of 3 min, a reaction period under aerobic conditions of 171 min, a settling period of 1 min, an effluent withdrawal period of 3 min and an idle period of 2 min. The hydraulic retention time (HRT) was always 0.25 days.

Table 2: Averaged values during the different operational stages of the SBR.

Period	Days (d)	COD (mg/L)	NH_4^+-N (mg/L)	COD/N (g/g)
I	12-70	500	100	5.0
II	71-160	500	200	2.5
III *	161-203	500	200	2.5
IV	204-210	500	33	15.0
V	211-246	500	72	7.0
VI	247-322	62.5	50	1.25
VII	323-342	0	50	0
VIII	343-392	0	100	0

* Removal of 5.25 g VSS from the reactor.

Analytical methods

The pH, dissolved oxygen, nitrate, nitrite, ammonia, volatile suspended solids (VSS) and SVI were determined according to the Standard Methods (APHA, 1985). Concentrations of chemical oxygen demand (COD) were determined by a modified method from Standard Methods (Soto *et al.*, 1989). Biomass density was measured using dextran blue according to Beun *et al.* (2002). Biomass was observed using a Stereomicroscope ZEISS 2000-C provided with a source of cold light KL 200.

Results and discussion

Nitrogen removal

During Period I the reactor was operated with a COD/N ratio of 5 g/g during 58 days. The ammonia concentration in the influent was 100 mg N/L and the NLR of 0.4 kg N/m^3·d (Figure 2). After 30 days of operation stable conditions were reached and the percentage of nitrogen removal was 48-54%. At this point the concentrations of inorganic nitrogen compounds in the effluent were around 12 mg NO_3^--N/L and 17 mg NO_2^--N /L. During the next 90 days the COD/N ratio applied was 2.5 g/g (Period II). The ammonia concentration in the influent was increased to 200 mg N/L and after the first 30 days stable conditions were reached and percentage of nitrogen removal was 33-41%. Concentrations in the effluent were around 30 mg NO_3^--N/L and 0 mg NO_2^--N /L. During Period III the fed COD/N ratio was 2.5 g/g but with a significant decrease of biomass concentration from Period II due to the removal of 5.25 g VSS to inoculate another reactor. The percentage of nitrogen removal was only 16-18% and concentrations in the effluent were around 5 mg NO_3^--N/L and 10 mg NO_2^- -N /L, respectively. During the Period IV concentrations of ammonia around 180 mg NH_4^+-N/L were present in the effluent and inhibition of the nitrification by free ammonia (FA) was expected (Anthonisen *et al.*, 1976). Yang *et al.* (2004) observed that in the presence of concentrations of more than 18 mg NH_3-N/L might cause an inhibition of the production of exopolymers and with a detriment of biomass aggregation and aerobic granules formation. This effect was not observed in the present study because the free ammonia concentration was kept between 4 and 8 mg NH_3-N/L. To avoid the inhibitory effect on the nitrifying activity due to free ammonia at high pH, the inlet ammonia concentration was decreased to 36 mg NH_4^+-N/L, causing a drop of the effluent concentration to 9 mg NH_4^+-N/L (Period IV). In this period the nitrogen removal was around 40%. The NLR was increased again to re-establish the nitrifying activity, however, no increase of the nitrifying rate was observed (Period VI). The concentration of the inlet COD was decreased during period VII in order to favour the growth of the nitrifying bacteria. The nitrifying activity was restored before 50 days of operation in these conditions. Finally, an autotrophic medium was fed (Period VII), obtaining a fully stable nitrification to nitrate. Nevertheless, when the NLR was doubled nitrite appeared as the main product (Period VIII). During these three last periods no nitrogen removal was observed meaning that almost 100 % of the fed ammonia nitrogen left the reactor in the form of ammonia, nitrite or nitrate.

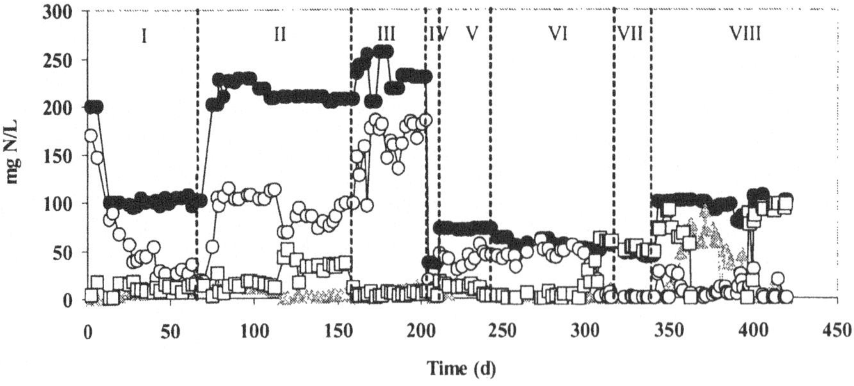

Figure 2: Ammonia concentration in the influent (●) and ammonia (○), nitrate (□) and nitrite (▲) concentrations in the effluent.

The organic loading rates (OLR) and the nitrogen loading rates (NLR) applied to the SBR during the experiment were in the range of 0 to 2 kg COD/m^3·d and 0.14 to 0.80 kg NH_4^+-N/m^3·d, respectively. The removal percentage of COD removal was almost constant (around 90%) while the nitrogen removal percentages were variable due to the changes of the operational conditions. The main processes for nitrogen removal in the SBR were nitrogen assimilation for biomass growth and nitrification-denitrification of ammonia. In order to discern between the percentages of nitrogen removal achieved by each of these mechanisms a nitrogen balance was calculated in the reactor to determine the amount of nitrogen used for growth. For each period the amount of biomass produced was estimated from the biomass increase in the reactor and the amount of biomass washed out in the effluent using equation [1]:

$$DW_P = \Delta Xr \cdot Vr + \bar{X}e \cdot Q \cdot \Delta t \quad \text{(g VSS)} \qquad [1]$$

Where: DW_P = amount of produced biomass (g VSS), ΔXr = change of biomass concentration during each period (g VSS/L), Vr = reactor volume (L), $\bar{X}e$ = averaged biomass concentration washed out in the effluent (g VSS/L), Q = flow rate (L/d) and Δt = length of each period (d). Considering a general composition of the biomass as $C_5H_7NO_2$ the averaged amount of nitrogen assimilated for biomass growth (DW_N) was calculated using equation [2] as:

$$DW_N = DW_P \cdot \frac{14 \text{ g-mol N}}{113 \text{ g-mol Biomass}} \quad \text{(g N)} \qquad [2]$$

Dividing the averaged amount of the nitrogen present in the produced biomass (DW_N) by the flow rate (Q) and the duration of each period (Δt) and referred to the inlet ammonia concentration, the percentage of nitrogen assimilated is calculated. The averaged percentages of nitrogen removed (N_R), calculated as the difference of the nitrogen amount in the feeding minus nitrogen in the effluent, and the expected percentages of assimilated nitrogen (N_A) are represented in Table 3. N_R and N_A were calculated related to the nitrogen amount in the feeding. Nitrogen removed by denitrification (N_D) was calculated by the difference between both of them. The fraction of nitrogen, removed by each mechanism, depended on the COD/N ratio of the influent in such a way that biomass assimilation can account for the removal of a large fraction of nitrogen when COD/N ratio in the influent to the SBR is high (Garrido *et al.*, 2001). The fact that the granules exhibited different denitrifying activities depends on the COD/N ratio. As occurred in biofilms O_2 is present only in the outer layers (Tijhuis *et al.*, 1994; van Loosdrecht *et al.*, 1995) and the inner layers, which were maintained under anoxic conditions, received the carbon source and nitrate to support the denitrification process. If the carbon source is not available enough (COD/N ratios low) the denitrification process is not completed and nitrite is produced as intermediate.

Biomass evolution

The concentration of the biomass increased from 3.5 to 9.0 g VSS/L (Figure 3) during Periods I and II when a constant inlet OLR of 2 kg COD /m^3·d was maintained. The biomass decrease observed in Period III is due to the removal of 5.25 g VSS. When the applied OLR was reduced to 0.25 kg COD /m^3·d (Period VI) and finally to zero (Periods VII and VIII) the biomass concentration decreased from 7.5 to 1.0 g VSS/L due to the heterotrophs washout from the system. The biomass concentrations during the addition of COD in the influent were in the range of those obtained by Beun *et al.* (2002) operating a granular SBR at an OLR of 2.5 kg COD /m^3·d. On the other hand, the solids concentration in the effluent was always lower than 0.1 g VSS/L, this value decreasing to 0.01 g VSS/L when no organic matter was applied to the reactor. This last value agrees with those obtained by Campos et al. (2000) who found solids concentrations in the effluent of 0.005-0.01 g VSS/L operating a nitrifying airlift reactor.

Table 3: Percentages of NH_4^+-N and NO_x^--N in the effluent and percentages of nitrogen removed (N_R), assimilated (N_A) and denitrified (N_D) referred to the inlet ammonia concentration.

Period	NH_4^+-N (%)	NO_x^--N (%)	N_R (%)	N_A (%)	N_D (%)
I	17-31	21-29	48-54	8	40-46
II	45-49	10-22	33-41	4	29-37
III	74-80	4-8	16-18	7	9-11
IV	24-27	20-22	53-55	39	14-16
V	48-68	22-34	10-18	15	0
VI	0	100	0	≈0	0
VII	0	100	0	≈0	0
VIII	3-9	91-97	0	≈0	0

The sludge volumetric index (SVI) trended to decrease during the first periods, achieving a stable value around 20 mL/g VSS (Figure 4). However during the autotrophic period there was a slight trend to increase. These low values indicate that the granules have a compact structure, which is related with a high settling velocity and a low requirement of volume after sedimentation (higher volumetric exchange ratio in the reactor). This agrees with Qin *et al.* (2004) who found that a decrease of the settling time caused a decrease of the SVI. On the other hand, Tay *et al.* (2001) also observed the influence of the air velocity in the SVI value. These authors found that high hydrodynamic shear forces seemed to stimulate the production of cellular polysaccharides, which play an important role in the formation of aerobic granules. Although not enough information is given in this case about the O_2 concentration corresponding to each stage, which could be a limiting substrate when the hydrodynamic shear forces where lower due to less aeration and produce granular sludge with bad settling properties due to this low O_2 concentrations.

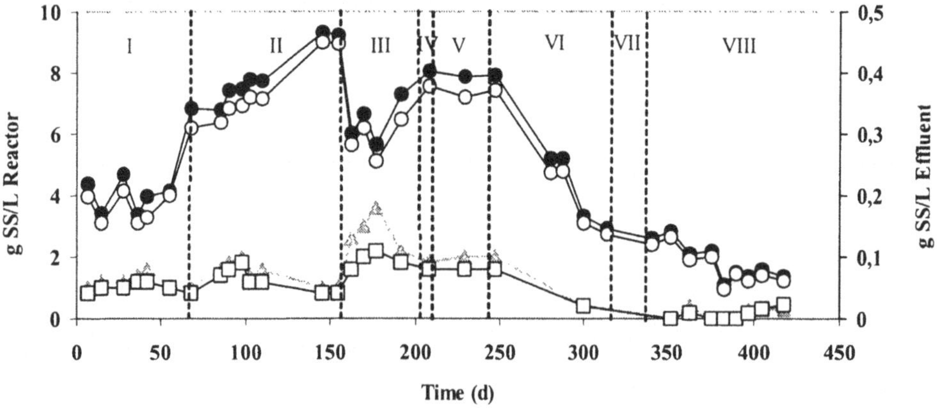

Figure 3: Concentration of VSS (O) and NVSS (●) in the system and concentration of VSS (□) and NVSS (▲) in the effluent.

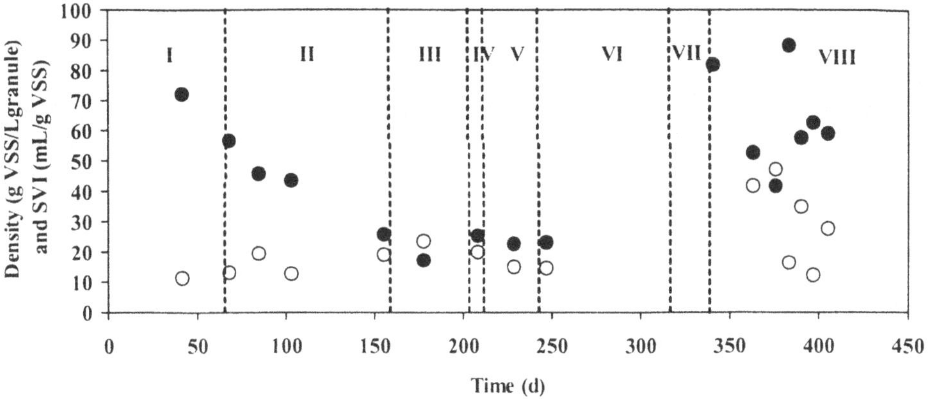

Figure 4: SVI (mL/g VSS) (●) and density (g VSS/L $_{granule}$) (O) of the biomass.

The density of the granules was between 10 and 30 g VSS/$L_{granule}$ and slightly increased to around 40 g VSS/$L_{granule}$ in period VIII when no organic carbon source was added and mainly nitrifiers were in the system. But due to the disappearance of the heterotrophs the granules broke apart and the density decreased again with the consequent increase of SVI (Figure 5).

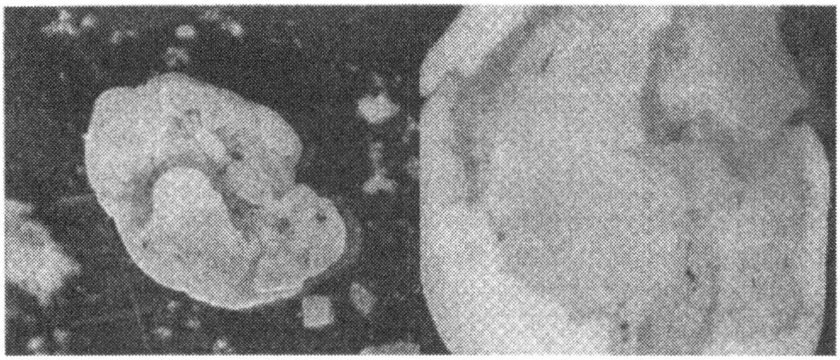

Figure 5: Autotrophic granules broke in pieces during period VIII.

Obtained nitrifying granules maintained their settling properties in spite of the change in the composition performed in Periods VII and VIII due to complete removal of organic carbon source from the influent. The structure of granules changed and they appeared as empty granules looking like the shells of the original granules. To obtain nitrifying granules from sludge from a municipal WWTP long periods of time can be required due to the slow growth of the autotrophic biomass and the possible inhibition of the process by FA. Nitrifying granules can be easily obtained using the fast growing heterotrophic bacteria to form the initial structure of the granules where the nitrifying microorganisms grow retained in between the heterotrophs.

Conclusions

An aerobic granular SBR was operated at different COD/N ratios achieving a removal of 90% for the organic matter and up to 55% for the ammonia nitrogen.

Removal of ammonia was carried out by both assimilation and simultaneous nitrification-denitrification. The predominance of each mechanism was related to the COD/N ratio in the feeding media.

Nitrifying granules are easily obtained by removing the organic carbon source from aerobic granules growing in a SBR where heterotrophic organic matter oxidation, denitrification and nitrification processes take place.

Acknowledgments

This work was funded by the Xunta de Galicia (Spain) through the GRAFAN project (Ref: PGIDT04TAM26500PR) and the Spanish CICYT which funded this research through the Oxanamon project (PPQ-2002-00771).

References

APHA-AWWA-WPCF. (1985) Standard Methods for examination of water and wastewater. 16th Ed. Washington.

Anthonisen A.C., Loehr R.C., Prakasam T.B.S. and Srinath E.G. (1976). Inhibition of nitrification by ammonia and nitrous acid. Journal WPCF, 48(5), S35-S52.

Beun J.J., Hendriks A., Van Loosdrecht M.C.M., Morgenroth E., Wilderer P.A. and Heijnen J.J. (1999). Aerobic granulation in a sequencing batch reactor. Water Research 33 (10), 2283-2290.

Beun J.J., Heijnen J.J. and van Loosdrecht (2001). N-removal in a granular sludge sequencing batch airlif reactor. Biotechnology and Bioengineering 75, 1, 82-92.

Beun J.J., Heijnen J.J. and van Loosdrecht (2002). Aerobic granulation in a sequnecing batch airlift reactor. Water Research, 36, 702-712.

Campos J.L., Méndez R. and Lema J.M. (2000). Operation of a nitrifying activated sludge airlift (NASA) without biomass carrier. Water Science and Technology, 41(4-5), 113-120.

de Beer D., van den Heuvel J.C. and Ottengraf S.P.P. (1993). Microelectrode measurements of the activity distribution in nitrifying bacterial aggregates. Appl. Environ. Microbiol., 59(2), 573-579.

Garrido J. M., Omil F., Arrojo B., Méndez R. and Lema J.M. (2001). Carbon and nitrogen removal from a wastewater of an industrial dairy laboratory with a coupled anaerobic filter-sequencing batch reactor system. Water Science and Technology 43, 315–321.

Qin L., Tay J.H. and Liu Y. (2004). Selection pressure is a driving force of aerobic granulation in sequencing batch reactors. Process biochemistry, 39, 579-584.

Smolders G.J.F., Klop J., van Loosdrecht M.C.M. and Heijnen J.J. (1995). A metabolic model of the biological phosphorus removal process. Effect of the sludge retention time. Biotechnology and Bioengineering 48, 222-233.

Soto M., Veiga M.C., Méndez R. and Lema J.M. (1989). Semi-micro COD determination method for high salinity wastewater. Environmental Technology Letters 10(5), 541-548.

Tay J.H., Lui Q.S. and Liu Y. (2001). The role of cellular polysaccharides in the formation and stability of aerobic granules. Letters in Applied Microbiology, 33, 222-226.

Tijhuis L., van Loosdrecht M.C.M. and Heijenen J.J. (1994). Formation and growth of heterotrophic aerobic biofilms on small suspended particles in airlift reactors. Biotechnology and Bioengineering 44, 595-608.

van Benthum W.A.J., van Loosdrecht M.C.M. and Heijnen J.J. (1997). Control of heterotrophic layer formation on nitrifying biofilms in a biofilm airlift suspension reactor. Biotechnology and Bioengineering 53, 397-405.

van Loosdrecht M.C.M., Tijuhuis L., Widieks A.M.S. and Heijnen J.J (1995). Population distribution in aerobic biofilms on small suspended particles. Water Science and Technology 31, 163–171.

Yang S.F., Tay J.H. and Liu Y. (2004). Inhibition of free ammonia to the formation of aerobic granules. Biochemical Engineering Journal, 17, 41-48.

Simultaneous phosphorus and nitrogen removal by aerobic granular sludge in single SBR system

X.M. Li, G.J. Yang, Q. Yang, G.M. Zeng, D.X. Liao, M.F. Hu and Y.M. Wu

Department of Environmental Science and Engineering, Hunan University, Yuelu Mountain, Changsha, 410082, P.R. China

Abstract Simultaneous biological phosphorus and nitrogen removal through enhanced anoxic phosphate uptake was investigated with aerobic granular sludge in an anaerobic/aerobic SBR system. Significant amounts of phosphorus-accumulating organisms (PAOs) capable of denitrification could be accumulated and enriched in the aerobic granular sludge coexisting with nitrifiers and nonPAO-denitrifiers under appropriate operational conditions. The ratio of the anoxic phosphate uptake to the aerobic phosphate uptake capacity reached 73.1% at 8~15mg/L PO_4^{3-}-P when dissolved oxygen concentration in the aerobic phase was controlled at 1~2mg/L, with 80~90min anaerobic phase followed by 240min aerobic phase during the SBR cycle. The anaerobic/aerobic SBR system showed a very stable phosphorus, nitrogen and organic carbon removal performance. The average removal rate for NH_4^+-N, total inorganic nitrogen (TIN), PO_4^{3-}-P and CH_3COO^--C reached 97.8%, 89.7%, 96.8% and 98.8%, respectively at the influent concentration of 8~15mg/L PO_4^{3-}-P, 25~50mg/L NH_4^+-N, 100~180 mg/L CH_3COO^--C, with the SBR system operated at 7.0g/L MLSS and 6.4g/L MLVSS respectively.

Keywords SBR; aerobic granular sludge; simultaneous phosphorus and nitrogen removal; denitrifying PAOs; anoxic phosphate uptake

Introduction

Eutrophication is a global problem, in which blooms of cyanobacteria and eukaryotic algae occur as a consequence of the breakdown of community homeostasis from continued pollution of oligotrophic aquatic bodies with nutrients like nitrogen (N) and phosphorus (P) at levels which exceed growth-limiting concentrations for these photosynthetic organisms (Winkler, 1984; Conley, 2000). Phosphorus and nitrogen removal from wastewater is an effective approach for prevention of eutrophication in closed water systems. In view of this, biological nutrient removal (BNR) processes have been widely used to treat nutrient materials in wastewater for the past several decades. Among BNR processes, the enhanced biological phosphorus removal (EBPR) process has been recognized as one of the most suitable techniques for avoiding the eutrophication problem since phosphorus acts as a limiting nutrient for algal growth.

Biological phosphorus removal from wastewater is based on the activity of phosphorus-accumulating organisms (PAOs). Under anaerobic conditions, PAOs take up readily biodegradable organic carbon substrates and store them as polyhydroxyalkanoates (PHAs). The energy to this anaerobic process is derived from the hydrolysis of intracellular polyphosphate and the glycolysis of glycogen followed by the release of ortho-phosphate to the bulk liquid (Mino et al., 1987). Under aerobic conditions, PAOs use the PHAs for generating energy for growth, glycogen synthesis, and phosphate uptake, where the PAOs take up phosphate more than released during the anaerobic phase (luxury uptake). It was assumed that PAOs could only use oxygen as an electron acceptor to accumulate phosphate under aerobic conditions in the early studies of biological nutrient removal processes (Wentzel et al., 1989). However, since then, there have been many reports claiming that a significant fraction of PAOs could take up phosphate using nitrate as an electron acceptor under anoxic conditions (Vlekke et al., 1988; Kerrn-Jesperson and Henze 1993; Kuba et al., 1993; Barker and Dold, 1996; Mino et al., 1998). It was hypothesized that biological phosphorus removal population comprises at least two groups: one group capable of utilizing either dissolved oxygen or nitrate as an electron acceptor (denitrifying PAOs), and the other group capable of utilizing only dissolved oxygen (aerobic PAOs) (Gerber et al., 1987; Kerrn-Jesperson and Henze, 1993; Meinhold et al., 1998). If the denitrifying PAOs take up and store phosphate using nitrate as electron acceptor, then the organic carbon substrate can be used simultaneously for both phosphorus and nitrogen removal. This is of significance since organic carbon content in most municipal wastewaters is too limited in order to achieve both phosphorus and nitrogen removal. Employing denitrifying PAOs in the biological nutrient processes also makes it possible to reduce sludge production and aeration demand (Copp and Dold, 1998).

Simultaneous phosphorus and nitrogen removal processes require an aerobic phase for nitrifying reaction to provide nitrate as an electron acceptor. The phosphorus and nitrogen removal processes can be classified into single- and two-sludge systems according to whether denitrifying PAOs coexist with nitrifiers or not. In the single-sludge system, the mixed sludge goes through all stages such as anaerobic, aerobic, and anoxic. Nitrifiers in the mixed sludge require a long aerobic period, but this is known to inhibit the growth and activity of denitrifying PAOs (Kuba et al., 1996a). Therefore, in single sludge system, it is of importance to make the conditions favourable for the denitrifying PAOs for the successful removal of phosphorus and nitrogen. The proportion of denitrifying PAOs could be increased in the SBR (from 11% to 64% of total PAOs) by introducing an anoxic phase into the middle of aerobic phase of the anaerobic-aerobic SBR-operation (Lee et al., 2001).

During the past few years, aerobic granular sludge has been developed in sequencing batch reactors (SBR) (Beun et al., 1999; Morgenroth et al., 1997). Compared to conventional activated sludge flocs, aerobic granular sludge has a regular, dense and strong physical structure, good settling ability, high biomass retention, and the ability to withstand shock-loading rate (Lin et al., 2003). In addition, as a result of the diffusion gradient of oxygen, aerobic zones, anoxic zones and anaerobic zones exist in aerobic granular sludge simultaneously (Zou et al., 2001). This offers conditions favourable for the denitrifying PAOs coexisting with nitrifiers.

In this work, the technical feasibility and stability of simultaneous phosphorus and nitrogen removal was investigated through aerobic granular sludge in a single sludge system by enhancing anoxic phosphate uptake. For this, appropriate dissolved oxygen concentration in the aerobic phase and operational conditions was adopted.

Material and methods

Sludge and wastewater

The seed sludge used was aerobic granular sludge with the ability of simultaneous nitrification and denitrification (SND), which was taken from a laboratory-scale aerobic/anaerobic SBR system and had a very compact and regular structure with a diameter of 0.5 to 1.0mm and a clear outer shape. Synthetic wastewater was used in this study, which consisted of sodium acetate (100~180mg/L CH_3COO^--C), NH_4Cl (25~50mg/L NH_4^+-N), KH_2PO_4 (8~15mg/L PO_4^{3-}-P), 94mg/L $MgSO_4 \cdot 7H_2O$, 21mg/L $CaCl_2$, 40mg/L $NaHCO_3$ and 1.6 ml of trace element solution (Smolders et al., 1994). Sodium acetate was used as the only organic substrate for it's ready biodegradability.

Reactor and operation

A sequencing batch reactor (SBR) with a working volume of 7L was used in this study. Two cycles were performed each day. Each cycle consisted of 6 phases, namely: influent (3mins), anaerobic (80~180mins), aerobic (240mins), settling (1.5mins), decanting (5.5mins) and idle (30mins). The sequence and duration of each cycle was pre-programmed into an electronic process controller. Temperature was controlled at 25±1℃ using a water circulation system. pH was controlled within 7~8 by using $NaHCO_3$ buffer. Nitrogen gas was introduced into the headspace of the reactor to prevent oxygen from entering into the system in the anaerobic stage. The reactor was equipped with slow stirrer to keep aerobic granular sludge in suspended state during the anaerobic stage. 4.0L synthetic wastewater was pumped into the reactor during the filling stage and the same amount of clarified supernatant was withdrawn from the reactor at the end of the settling stage.

Table 1: Operating conditions in the respective phase

Experimental phase	Duration (d)	SBR-cycle mode (mins) anaerobic/aerobic	Influent composition (mg/L) NH_4^+-N/PO_4^{3-}-P/CH_3COO^--C	DO (mg/L) anaerobic/aerobic	pH
1	30	120 / 240	25~30 / 8~12 / 100~120	<0.1 / 3	7~8
2	60	180,120, 90 / 240	25~35 / 8~12 / 100~140	<0.1 / 3, 2, 1	7~8
3	120	80~90 / 240	25~50 / 8~15 / 100~180	<0.1 / 1~2	7~8

There were 3 phases during whole research (Table 1). In phase 1, the aerobic granular sludge with the function of simultaneous phosphorus and nitrogen removal was obtained and sludge retention time (SRT) was maintained at 25d. In phase 2, the good performance of phosphorus, nitrogen and organic carbon removal was achieved after 2 months of operation, during which the operating conditions were continuously modified. By continuous exploration, the final operational conditions were determined as 80~90min anaerobic, 240min aerobic, 1~2mg/L DO during aerobic stage, 7.0g/L MLSS, 6.4g/L MLVSS and 20d SRT. And under these conditions in phase 3, the SBR system was continuously operated over 4 months for system ability and stability investigations. During this phase, the compositions of influent wastewater were changed: NH_4^+-N 25~50mg/L, PO_4^{3-}-P 8~15 mg/L. In order to maintain high removal efficiency of nitrogen and phosphorus, CH_3COO^--C was changed accordingly from 100 to 180mg/L to keep C/N ratio at a level of 4 approximately.

Batch experiment

After confirming the steady-state, batch experiments were carried out at regular intervals to investigate the proportion of denitrifying PAOs to the total PAOs and anoxic phosphate uptake in phase 3. The sludge was divided equally into two parts after the end of phosphorus release during the anaerobic phase was observed. One part of the sludge was exposed to anoxic conditions, and the other part was exposed to aerobic conditions. Under anoxic conditions, a dose of 50mg/L NO_3^--N was added into the reactor at the starting point of anoxic phase. After this initial dose, the nitrate was continuously added in order to supplement consumed nitrate. The amounts of phosphate uptake were measured under these conditions. During the anoxic phase, nitrogen gas was purged above the liquid surface to secure strictly anoxic conditions. The pH was also strictly controlled at 7.0 by addition of 0.5N HCl or 0.5N NaOH to avoid phosphate precipitation throughout the experiment.

Analytical methods

For monitoring the performance of the reactors, samples were retrieved from the reactor at regular intervals. NH_4^+-N, SV, SVI, MLSS and MLVSS were determined according to APHA Standard Methods (1989). CH_3COO^--C, PO_4^{3-}-P, NO_3^--N, NO_2^--N were analyzed using ion-chromatography (Dionex DX-120,USA) routinely. pH was measured by pH meter automatically. DO concentration was measured by an oxygen sensor (CellOx325). Total inorganic nitrogen (TIN) was calculated as below: TIN=$[NH_4^+$-N]+[NO_2^--N]+[NO_3^--N].

Results and discussion

Removal efficiency of nitrogen

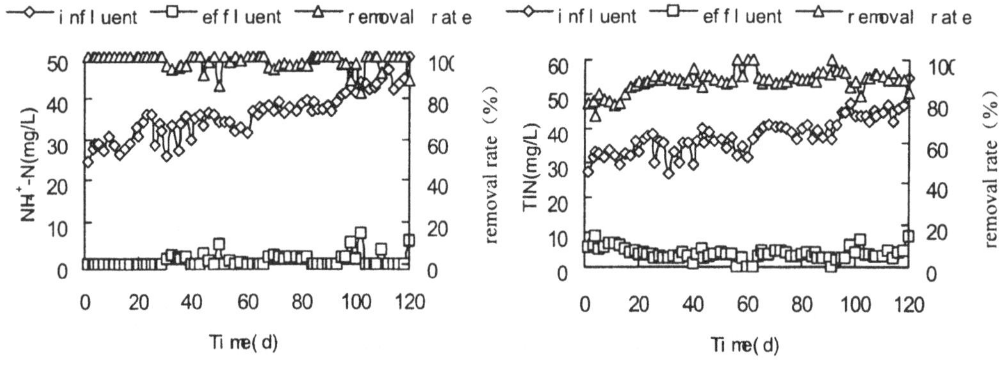

Figure 1: Removal rate of nitrogen by aerobic granular sludge in phase 3

Figure 1 shows the concentrations and removal rate for NH_4^+-N and TIN in phase 3. As can be seen in Figure 1, when the concentrations of NH_4^+-N and TIN in the influent were within 25~50mg/L and 26.7~54.9mg/L respectively, the effluent concentrations of them could be reduced to 0~5.7mg/L and 0~8.9mg/L respectively after treated by the SBR system. The NH_4^+-N removal rate ranged from 88.6% to almost 100%, and the average removal efficiency reached 97.8%. The TIN removal rate ranged from 73.2% to almost 100% and the average removal rate reached 89.7%. Those results suggested that simultaneous nitrification and denitrification could happen efficiently in the process when dissolved oxygen concentration was controlled at 1~2 mg/L during aerobic stage, which indicates that aerobic granular sludge has an excellent performance in nitrogen removal.

Removal efficiency of phosphorus

Figure 2 shows the concentration and removal efficiency of PO_4^{3-}-P in phase 3. As shown in Figure 2, the influent PO_4^{3-}-P concentration was within 8~15mg/L and the concentration in the effluent could be reduced to 0~2.5mg/L after treated by the SBR system. The PO_4^{3-}-P removal rate was as high as 82 to 100%, and the average removal efficiency reached 96.8%

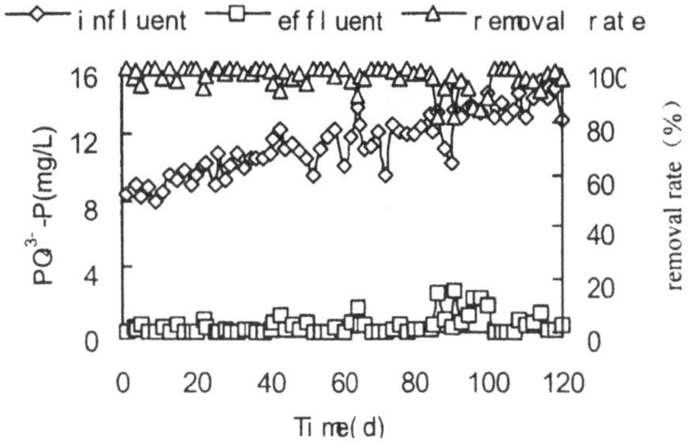

Figure 2: Removal rate of phosphorus by aerobic granular sludge in phase 3

during the whole process, which demonstrated the good ability of phosphorus removal by aerobic granular sludge.

Removal efficiency of CH_3COO^--C

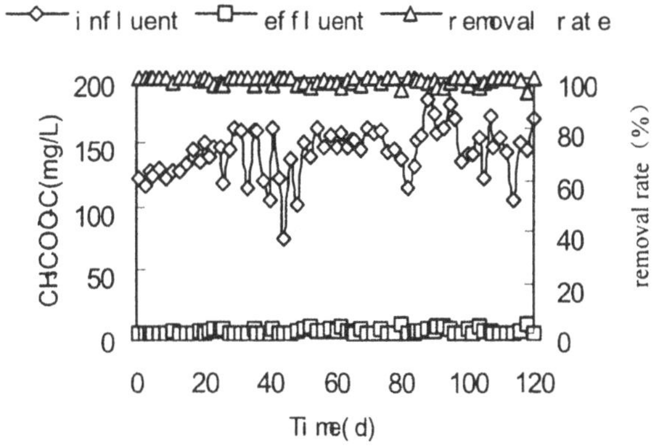

Figure 3: Removal rate of CH_3COO^--C by aerobic granular sludge in phase 3

As denitrification and phosphorus removal process both consume organic substance, the removal of nitrogen, phosphorus and organic substance, therefore can be realized simultaneously. Figure 3 shows the concentration and removal efficiency of CH_3COO^--C in phase 3. As shown in Figure 3, the influent CH_3COO^--C concentration was about 100~180mg/L and the concentration of CH_3COO^--C in the effluent declined to 0~8.4mg/L after treated by the SBR system. The CH_3COO^--C removal rate was as high as 94.2 to 100% and the average removal rate reached 98.8%, which illustrated the high performance in the degradation of a readily degradable organic carbon source by the aerobic granular sludge SBR system.

The nitrogen and phosphorus removal process throughout one cycle

Figure 4 shows the profiles of NH_4^+-N, CH_3COO^--C, PO_4^{3-}-P, NO_2^--N and NO_3^--N in a typical cycle of the SBR process in phase 3. As shown in Figure 4, when the concentrations of NH_4^+-N, CH_3COO^--C and PO_4^{3-}-P in the influent were at 36.1mg/L, 130.2mg/L and 15.4mg/L, respectively, carbon substrate was rapidly depleted and phosphate was sharply released after the anaerobic phase started. And at the end of the anaerobic phase, the concentrations of phosphate increased to around 131.4mg/L and the concentrations of CH_3COO^--C decreased to around 5.6mg/L. This was a result that PAOs took up organic carbon substrates and stored them as PHB and released ortho-phosphate to the bulk liquid. The concentration of ammonium varied lightly in the anaerobic phase.

Although the ammonium concentration declined continuously in the subsequent aerobic phase, the nitrate and nitrite concentration within this process kept a lower level (<1mg/L). At the end of aeration, ammonium and nitrite could not be detected and the concentration of nitrate remained at only 3.5mg/L. Those results indicate that nitrate and nitrite produced by nitrification were reduced to nitrogen gas by denitrifying bacteria (including denitrifying PAOs and nonPAO-denitrifiers) in the aerobic granular sludge and support the previous assumption of simultaneous nitrification and denitrification.

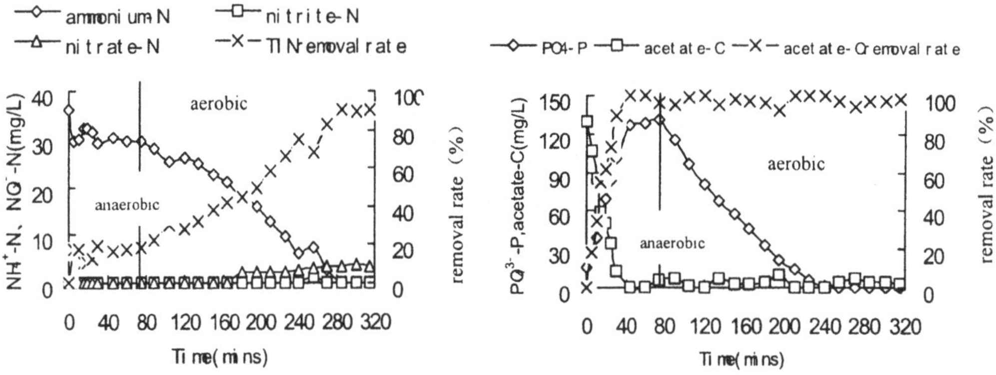

Figure 4: The nitrogen and phosphorus removal process throughout one cycle in phase 3

In addition, as can be seen from Fig. 4, the released phosphorus during the anaerobic phase was taken up luxuriously in the aerobic phase and phosphorus could not be detected at the end of aeration. It could be hypothesized there was not only the occurrence of aerobic phosphate uptake, there might also be the occurrence of denitrifying dephosphatation simultaneously during the aerobic phase. This was because denitrifying PAOs existed in the aerobic granular sludge, which would be demonstrated by the batch experiments results (Fig.5). Meanwhile, nitrate produced by nitrification could be provided as the electron acceptor for denitrifying PAOs. Furthermore, as a result of the diffusion gradient of oxygen, anoxic zones existed in the aerobic granular sludge when the mean size of granules was 0.5mm and DO concentration was controlled at 1~2mg/L in the bulk liquid, which, therefore, provided a suitable condition for denitrifying dephosphatation.

As shown in Figure 4, the concentration of CH_3COO^--C decreased lightly from 5.6mg/L at the end of the anaerobic phase to 2.7mg/L at the end of the aerobic phase. This result clearly indicated that intracellular PHB was used as the energy sources for denitrification and phosphate uptake in the aerobic phase, which would be beneficial for both phosphorus and nitrogen removal if the available amount of carbon source in wastewater is limited.

During this cycle, the removal rate for CH_3COO^--C, TIN and phosphorus reached 97.9%, 90.3% and almost 100%, respectively.

The proportion of denitrifying PAOs to the total PAOs of granular sludge

It has been shown that the contribution of phosphate removal by denitrifying PAOs to the total phosphorus removal could be calculated from the ratio of the anoxic phosphate uptake rate to the aerobic phosphate uptake rate (Wachtmeister et al., 1997). This is based on the fact that denitrifying PAOs can take up phosphate at nearly the same rate under both aerobic and anoxic conditions, whereas aerobic PAOs are inactive under anoxic conditions.

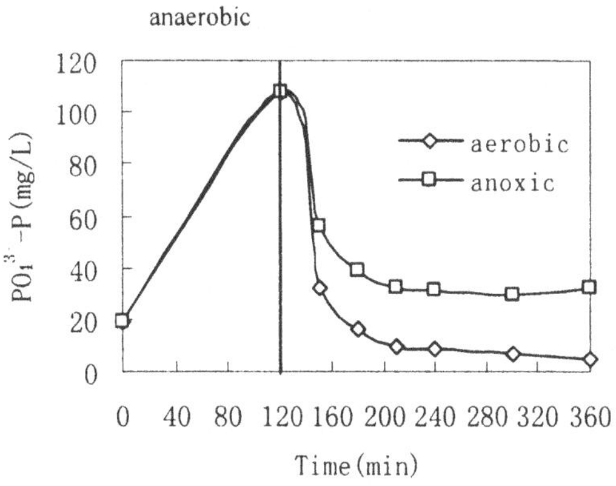

Figure 5: Phosphorus uptake test under aerobic and anoxic conditions in phase 3

Figure 5 shows the phosphate release and uptake by the aerobic granular sludge under different conditions, which presents data collected from the batch experiments results in phase 3. As can be seen from Fig.5, the concentration of phosphorus reached 108.1mg/L at the end of the anaerobic phase. After being exposed to anoxic and aerobic conditions for 4h, the concentration of phosphorus decreased to 32.2mg/L and 4.7mg/L, respectively. The rates of phosphate uptake under anoxic and aerobic conditions were 18.9mg/(L·h) and 25.9mg/(L·h), respectively. Since the activity of anoxic phosphate uptake was approximately 73.1% of the aerobic activity, the proportion of denitrifying PAOs was estimated to be 73.1% of total PAOs in the aerobic granular sludge. Therefore the conclusion can be made that denitrifying PAOs were enriched in the aerobic granular sludge coexisting with nitrifiers and nonPAO-denitrifiers.

Conclusion

Although longer aerobic time (240min aerobic) was required for the nitrification in the anaerobic/aerobic SBR, significant amounts of PAOs capable of denitrification could be accumulated and enriched in the aerobic granular sludge coexisting with nitrifiers and nonPAO-denitrifiers under appropriate operational conditions: 80~90min anaerobic, 240min aerobic, 1~2mg/L DO during aerobic stage, 7.0g/L MLSS, 6.4g/L MLVSS and 20d SRT. The proportion of denitrifying PAOs was estimated to be 73.1% of total PAOs in the aerobic granular sludge in this research. The anaerobic/aerobic SBR system showed a very stable phosphorus, nitrogen and organic carbon removal performance. Average removal efficiencies of ammonium, total inorganic nitrogen, phosphorus and CH_3COO^--C were 97.8%, 89.7%, 96.8% and 98.8%, respectively.

References

APHA (1989) Standard methods for the Examination of Water and Wastewater. 17thed. American Public Health Association, Washington, DC.

Barker, P.S. and Dold, P.L. (1996) Denitrification behaviour in biological phosphorus removal activated sludge systems–review paper. *Wat. Res.* 30(4), 769–780.

Beun, J.J., Hendriks, A., van Loosdrecht, M.C.M., Morgenroth, E., Wilderer, P.A., Heijnen, J.J. (1999)Aerobic granulation in a sequencing batch reactor. *Wat. Res.* 33 (10), 2283–2290.

Conley, D.J. (2000) Biogeochemical nutrient cycles and nutrient management strategies. Hydrobiologia 410, 87–96.

Copp, J.B. and Dold, P. (1998) Comparing sludge production under aerobic and anoxic conditions. *IAWQ 19th Biennial International Conference*, pp. 268–275.

Gerber, A., de Villiers, R.H., Mostert, E.S. and van Riet, C.J. (1987) The phenomenon of simultaneous phosphate uptake and release, and its importance in biological nutrient removal. In Biological Phosphate Removal from Wastewaters, ed. R. Ramadori, Pergamon Press, Oxford.

Kerrn-Jesperson, J.P. and Henze, M. (1993) Biological phosphorus uptake under anoxic and aerobic conditions. *Wat. Res.* 27(4), 617–624.

Kuba, T., van Loosdrecht, M.C.M. and Heijnen, J.J. (1996a) Effect of cyclic oxygen exposure on the activity of denitrifying phosphorus removing bacteria. *Wat. Sci.Tech.* 34(1–2), 33–40.

Kuba, T., Smolders, G.J.F., van Loosdrecht, M.C.M. and Heijnen, J.J. (1993) Biological phosphorus removal from wastewater by anaerobic– anoxic sequencing batch reactor. *Wat. Sci. Tech.* 27, 241–252.

Lee, D.S., Jeon, C.O. and Park, J.M. (2001) Biological nitrogen removal with enhanced phosphate uptake in a sequencing batch reactor using single sludge system. *Wat. Res* .35(16), 3968–3976.

Lin, Y.M., Liu, Y. and Tay, J.H. (2003) Development and characteristics of phosphorus-accumulating microbial granules in sequencing batch reactors. *Appl. Microbiol. Biotechnol.* 62, 430–435.

Meinhold, J., Pedersen, H., Arnold, E., Isaacs, S. and Henze, M. (1998) Effect of continuous addition of an organic substrate to the anoxic phase on biological phosphorus removal. *Wat. Sci. Tech.* 38(1), 97–105.

Mino, T., Arun, V., Tsuzuki, Y., Matsuo, T. (1987) Effect of phosphorus accmulation on acetate metabolism in the biological phosphours removal process. Proceedings of the IAWPRC InternationalConferenc e on Biological Phosphate Removalfrom Wastewaters, Rome, Adv. Water Pollut. Cont., ed. R. Ramadori, pp. 27–38. Pergamon Press, Oxford.

Mino, T., van Loosdrecht, M.C.M. and Heijnen, J.J. (1998) Microbiology and biochemistry of the enhaned biological phosphate removal process. *Wat. Res.* 32(1), 3193–3207.

Morgenroth, E., Sherden, T., van Loosdrecht, M.C.M., Heijnen, J.J., Wilderer, P.A. (1997) Aerobic granular sludge in a sequencing batch reactor .*Wat. Res.* 31(12), 3191–3194.

Smolders, G.J.F., van der Meji, J., van Loosdrecht, M.C..M., and Heijnen, J.J. (1994) Stoichiometric model of the aerobic metabolism of the biological phosphorus removal process. *Biotechnol. Bioeng.* 44, 837–848.

Vlekke, G.J.F., Comeau, Y. and Oldham, W.K. (1988) Biological phosphate removal from wastewater with oxygen or nitrate in sequencing batchreactors. *Environ.Technol. Lett.* 9, 791–796.

Winkler, M. (1984) Biological control of nitrogenous pollution in wastewater. In: Topics in Enzyme and Fermentation Biotechnology Vol.8(Wiseman, A., Ed.), pp.31–125. Ellis Horwood, Chichester.

Wentzel, M.C., Ekama, G.A., Loewenthal, R.E., Dold, P.L. and Marais, G.v.R. (1989) Enhanced polyphosphate organisms cultures in activated sludge systems. Part II: experimental behaviour. *Wat. S. A.* 15, 71–81.

Wachtmeister, A., Kuba, T., van Loosdrecht, M.C.M. and Heijnen, J.J. (1997) A sludge characterization assay for aerobic and denitrifying phosphorus removing sludge. *Wat. Res.* 31(3), 471–478.

Zou, L.P., Liu, X.D., Wang, B.Z., and Fan, Y.Z. (2001) The influence of ecological factors on simultaneous nitrification and denitrification in MBR. *Environ. Sci.* (China) 22 (4), 51–55.

Bio-P removal profile of aerobic granular activated sludge from an anaerobic/aerobic SBR system

Jianrong Zhu*, Chunxin Liu*, and P. A. Wilderer**

* Dept. of Environmental Engineering, Tsinghua University, Beijing 100084, P.R. China
** Institute of Advanced Studies on Sustainability, European Academy of Sciences and Arts, c/o Wasserguete, TU Muenchen, Am Coulombwall, 85748 Garching, Germany

Abstract The bio-P removal profile of granular activated sludge developed in a lab-scale SBR was examined before, during, and after completion of sludge granulation. The P-removal efficiency improved with the development of the granulation process. As long as flocculent sludge prevailed, bio-P removal only reached about 60% but gradually improved to 67-87% with the increase of the fraction of granular activated sludge. The P removal efficiency eventually reached a maximum 99% after completion of granular sludge formation in the reactor. Bio-P release occurred during the no-mix phase of the SBR cycle, mainly during the fill phase. The maximum P concentration after release was 5 times higher than the P concentration in the feed, and the maximum release rate was 2.4 mg P min^{-1} L^{-1}. Bio-P uptake proceeded rapidly during the initial stage of the aerobic react phase. The maximum bio-P uptake rate was about 2.2 mg P min^{-1} L^{-1}, which was very close to the maximum P release rate. The P content in granular sludge went up to almost 9%, and its variable profile during one cycle is evidence that a bio-P process was the major removal mechanism. These results demonstrated that a bio-P removal can be effective in a granular activated sludge SBR.

Keywords bio-P removal, aerobic granular activated sludge, SBR technology, wastewater treatment

Introduction

The mechanisms leading to enhanced biological phosphorus removal (EBPR) have been studied by a multitude of researchers (Liu *et al.*, 1997; Seviour *et al.*, 2003). In principle, polyphosphate accumulating microorganisms (PAO) are understood to utilize intracellular polyphosphate to provide the energy needed to sequester, under anaerobic conditions, readily biodegradable COD and store it mainly as polyhydroxybutyrate (PHB). Under aerobic conditions, the storage products are metabolized and the polyphosphate pool is regenerated through uptake and polymerization of ortho-phosphate. Release and uptake of ortho-phosphate are the characteristic mechanisms performed by PAOs and the key for success of EBPR plants.

Many attempts have been made to implement EBPR in continuous flow activated sludge plants to use the specific metabolic function of PAOs (Stephens and Stensel, 1998; Carucci *et al.*, 1999; Sudiana *et al.*, 1999). Sequencing Batch Reactors (SBRs) have also been widely applied in small scale and large scale (Wilderer *et al.*, 2001; Irvine *et al.,* 1997) to achieve EBPR. SBRs were also successfully used to grow and enrich granular activated sludge (Zhu and Liu, 1999; Beun *et al.*, 1999; Zhu and Wilderer, 2003). However, little is known about the possibility to enrich PAOs in granular activated sludge and about performance and long term stability of the respective reactors. Experimental trials were conducted to answer those questions.

Materials and Methods

Experimental reactors

Aerobic granular activated sludge was grown in a laboratory scale SBR designed as a bubble column. The reactor was made of glass having a total volume of 10 L, a working volume of 8 L, and a depth of 100 cm after completion of fill. Activated sludge from a municipal wastewater treatment plant was chosen as inoculum, and synthetic wastewater containing glucose as the main carbon source served as feed. In the feed, the concentration of organic substrates was 450 mg COD L^{-1}, and the phosphorous concentration was adjusted to10.5 mg L^{-1}. The SBR was operated at ambient temperature (approximately 20 °C) with no pH control. The sludge concentration achieved in the reactor was an average 11 g L^{-1}, the sludge age was set to 30 – 40 days, the volumetric exchange ratio to 75%, and the cycle time to 360 minutes. The reactor was fed from the top during a no-mix fill phase of 15 minutes. After completion of fill the aerator was turned on for 5 minutes to evenly distribute the wastewater introduced into the reactor. During the following 95 min anaerobic/anoxic react phase, the reactor remained unmixed and unaerated to achieve denitrification and to stimulate P release. The aerobic react phase that followed lasted 210 minutes. After the aerator was turned off the sludge was allowed to settle for 20 minutes. The supernatant was then removed during a decant phase of 15 minutes.

Figure 1 provides an overview of the granules which developed in the reactor. The size of the granules varied between 1 and 2 mm. Over a period of one year, stable conditions with respect to size and performance of the granules could be observed.

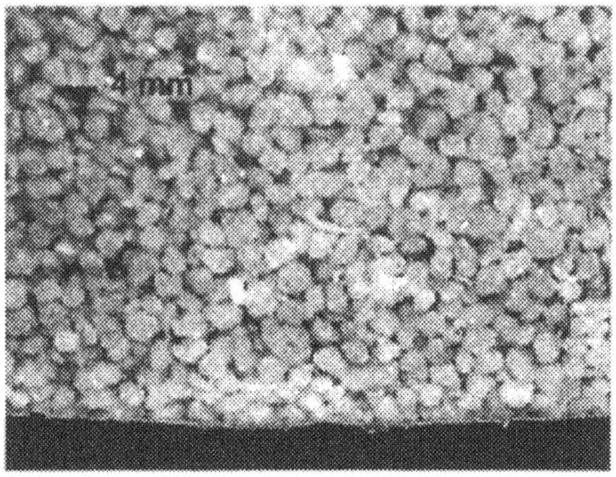

Figure 1: Typical morphology of mature granules

Analysis

- Parameters such as MLSS, MLVSS and SV were analyzed using standard methods (*Standard Methods*, 1998). COD, NH_4-N, NO_3-N and PO_3^{3-} were analyzed by the Dr. Lange test.
- To characterize the metabolic activity of the granules the OCR (oxygen consumption rate) was determined according to standard methods (*Standard Methods*, 1998) using a DO meter (WTW OR 156) and a Hydro data acquisition unit (Fluka company).

Results

Bio-P removal at various stages of granulation

During the first experimental phase I, mostly activated sludge flocs were present in the reactor with very few granules. As the values presented in Table 1 demonstrate, it took about 40 days to get bio-P removal established in the reactor. Eventually, the P removal efficiency reached 60%, while the efficiency of COD and nitrogen removal was, on average, 93% and 64%, respectively. The oxygen consumption rate varied in the range of 0.6 and 0.8 mg O_2 min^{-1} L^{-1}

Table 1: Bio-P removal of flocculent sludge

Time (d)	Influent P (mg L^{-1})	Effluent P (mg L^{-1})	P removal efficiency (%)
12	10.10	7.55	24.8
19	10.86	8.63	21.4
32	9.77	7.14	27.0
35	9.62	7.72	22.0
42	8.99	3.58	64.2
46	14.70	5.67	60.0
47	9.42	3.62	62.0

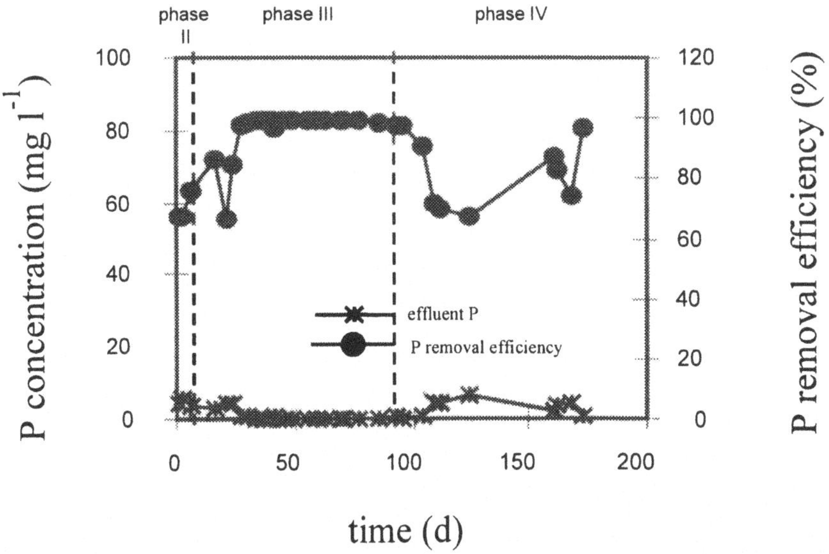

Figure 2: Course of bio-P removal during experimental phase II, III and IV

While continuing the experiment, the fraction of granules in the activated sludge gradually increased. During the experimental phase II, the sludge was partially granulated. During phase III, complete granulation was achieved (see Figure 1), and the bio-P removal efficiency reached 99% on average. During the subsequent experimental phase IV, the performance of the activated sludge deteriorated but recovered (Fig. 2). The drop in performance was mainly caused by a significant removal of granular sludge for experimental purposes. Approximately 18% of the granular sludge was taken from the reactor and used for several batch experiments. At the same time, nitrogen gas was blown into the reactor during the anoxic/anaerobic react phase of the SBR cycle to study the effects of mixing on bio-P release. Obviously, the change in the mixing regime and the increase of sludge loading were the reasons for the observed instabilities. However, the bio-P removal capacity of the sludge recovered after some time and finally reached almost the same level as before, namely 97%.

Evolution of the phosphorous concentration during the SBR cycle

The evolution of the phosphorous concentration during the SBR cycle is graphically represented in Figure 3. The concentration profiles shown were taken 45 and 71 days after commencement of experimental phase II, i.e. in the middle of experimental phase III. Both curves follow basically the same pattern.

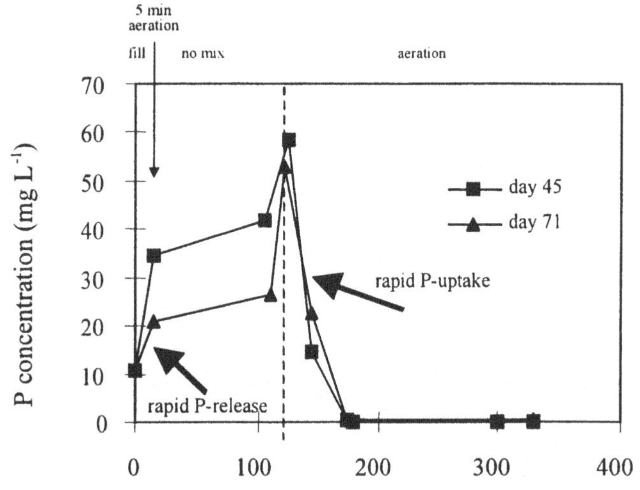

Figure 3: Phosphorous release and phosphorous uptake during selected SBR cycles

A rapid bio-P release occurred during the first 20 minutes of the cycle, particularly on day 45. As mentioned above, after completion of the fill phase, the reactor content was mixed for 5 minutes by means of weak bubble aeration. Thereafter, no mixing was provided, and the sludge was allowed to settle. Not surprisingly, the P concentration in the supernatant did not change very much during this time. The concentration significantly increased to 24 and 35 mg L^{-1}, respectively, indicating that P was released in large quantities during the anaerobic react phase. Actually, the concentration of phosphorus increased by 2.5-3.5 times during the anaerobic react phase of the SBR cycle. On average, the P-release rate during the fill phase was 1.4 mg P min^{-1} L^{-1}, and the maximum observed P release rate was 2.4 mg min^{-1} L^{-1}.

Within the first 30 minutes of the aerobic react phase the phosphorous concentration dropped to almost zero. With reference to the examples given in Figure 3, the maximum observed uptake rate was 2.2 and 1.3 mg P min^{-1} L^{-1}, respectively. It is worth noting that the maximum uptake rate was very similar in value to maximum release rate (2.2 to 2.4 mg P min^{-1} L^{-1}). These results are clear evidence of the high bio-P removal potential that can be established in granular activated sludge.

P release and accumulation in the granular sludge

The P content in granular activated sludge was measured during selected SBR cycles in order to explore the prevailing P removal mechanisms. A typical result is shown in Figure 4. The P content of the sludge decreased from 85 to 82 (mg P g SS^{-1}) during the fill phase but apparently remained relatively stable during the anaerobic react phase. Most probably, this result is an artefact caused by the lack of mixing. The first sample taken after the beginning of the aerobic react phase may better reflect the true situation. The P content of the sludge obviously had dropped during the anaerobic react phase to 77 mg P g SS^{-1}.

During the first 30 minutes in the aerobic react phase, the P content of sludge rose almost to the same level observed at the beginning of the cycle, which reflects the P uptake that happened during the same time interval as described in Figure 3. This is clear proof that the observed removal of P was the result of storage of phosphorous in the bacterial cell in the form of polyphosphate, and not the result P precipitation.

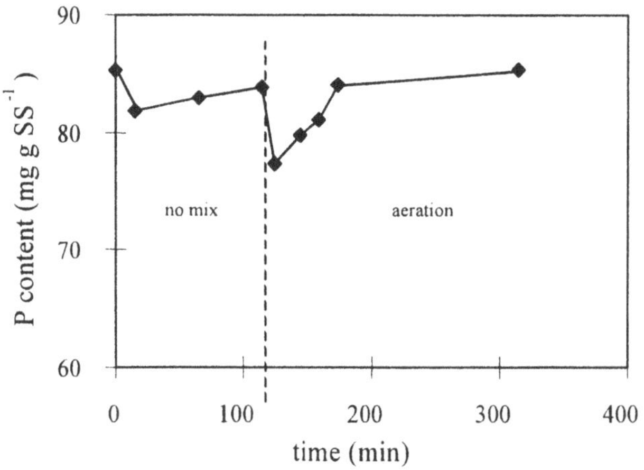

Figure 4: The P content profile of granular activated sludge during a typical cycle

Bio-P release during the anaerobic phase under mixing condition

To determine the bio-P release process in detail, P release was also examined when mixing was provided during the fill and the subsequent anaerobic react phase of the SBR cycle. To avoid transfer of oxygen into the bulk liquid, mixing was achieved by blowing nitrogen gas bubbles into the reactor. It was expected that under these conditions the released phosphorus would readily be distributed in the bulk liquid.

A typical example of the results achieved is presented in Figure 5. It was found that during the first 40 minutes the bio-P release rate was significantly higher than during the remaining period. Approximately 24 mg L^{-1} of phosphorous was released during this first phase of P release. The rate of release was about 1 mg P L^{-1} min^{-1}, a value which is lower than the release rates reported above but still in the same order of magnitude. This result is also confirmed by the results shown in Figure 6. Here, no mixing was provided, but the sludge was kept settled. Samples were taken during the anaerobic react phase from the bottom of the reactor (Fig.6).

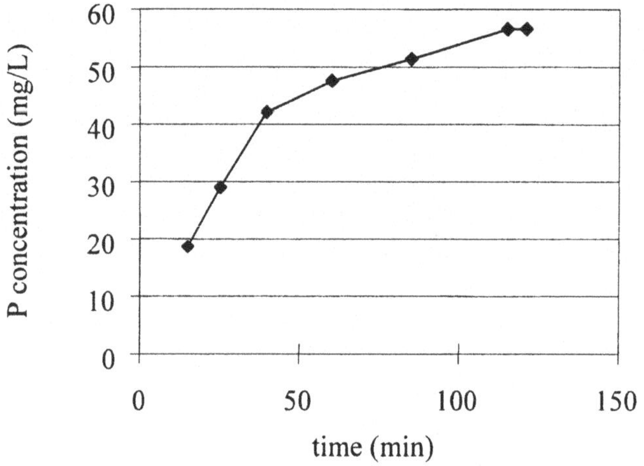

Figure 5: Bio-P release of granular activated sludge mixed by means of nitrogen gas dispersion

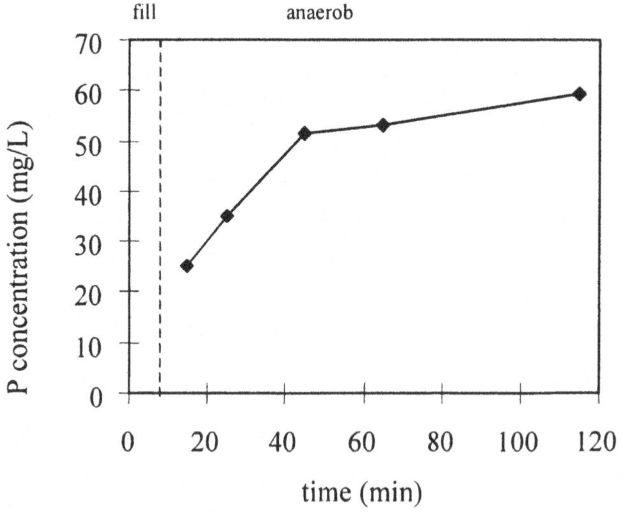

Figure 6: Bio-P release profile in the interstitial water within the settled sludge

Discussion

Many different reactor configurations have been proposed and implemented to achieve bio-P removal. In any case, enrichment of polyphosphate accumulating microorganisms (PAOs) in the activated sludge is the key of success. PAOs can be enriched by circulating the activated sludge through anaerobic and aerobic zones of a continuous flow activated sludge reactor system. SBR systems provide the possibility of alternating anaerobic/aerobic conditions along a time axis.

The experimental result described above clearly demonstrated that it is possible to establish good bio-P removal in SBR systems. As granular sludge developed in the SBR, the bio-P removal efficiency improved. The authors of this paper assume that the specific architecture of the granules, the corresponding mass transport processes, and the resulting zonation within the granules were the reasons for the improvement of the bio-P performance with granulation.

There have been several reports claiming that the bio-P removal cannot be accomplished in a reactor fed with glucose as the major carbon source because of preferential growth of so-called "G-bacteria", which do not accumulate polyphosphate (Liu *et al.*, 1996; Kong *et al.*, 2002). The experiments described in this paper do not support this assumption. The reason could be attributed to a synergistic association of PAOs with another group of bacteria in particular zones of the granules. It should be realized that during each SBR cycle, organic substrates, dissolved oxygen, ortho-phosphate, and nitrate are allowed to penetrate to varying depths of the granules, depending on the way the SBR is operated. Therefore, a very complex and time-dependent situation exists in the SBR, and in the granules in particular, which is not yet well understood. Further research is needed to gain a better understanding of the metabolic and transportation processes that are active within the granules in an SBR. Based on the observation made during our experimental trials, we assume that the growth conditions provided in the SBR are particularly favourable for the PAO population and finally stimulate bio-P release and uptake.

Conclusions

The experimental results described in this paper lead to the following conclusions:

- It is readily possible to develop granular activated sludge in an SBR fed with glucose as the major carbon source and operated according to an anaerobic react – aerobic react protocol.
- The ortho-phosphate introduced into the SBR during the fill phase was effectively removed from the bulk liquid, transferred into the granular sludge, and removed from the reactor by means of sludge wasting.
- Release of ortho-phosphate into the bulk liquid during the anaerobic react phase, particularly during the fill phase, and uptake of ortho-phosphate during the first 30 minutes in the aerobic react phase are strong indications that bio-P removal processes are responsible for the observed P removal.
- Systematic change of the P content in the granular activated sludge support the assumption that bio-P removal took place in the SBR.
- The maximum P release rate observed was 2.4 mg P min^{-1} L^{-1}, and the maximum P uptake rate was about 2.2 mg P min^{-1} L^{-1}.

- Systematic change of the penetration depth of organic substrates, dissolved oxygen, ortho-phosphate, and nitrate during the SBR cycle is assumed to be decisive for the stable performance of the SBR under investigation.

Acknowledgements

This research was financially supported by the Alexander von Humboldt Foundation, Germany, and by the Tsinghua University Foundation, P. R. China.

References

Beun J.J., Hendriks A., van Loosdrecht M.C.M, Morgenroth E., Wilderer P.A. and Heijnen J.J. (1999). Aerobic granulation in a sequencing batch reactor. *Wat. Res.*, **33** (10), 2283-2290.

Carucci A., Lindrea K., Majone M., and Ramadori R. (1999). Different mechanisms for the anaerobic storage of organic substrate and their effect on enhanced biological phosphorus removal (EBPR). *Wat. Sci. Tech.*, **39** (6), 21-28.

Irvine R.L., Wilderer P. A., and Flemming H.-C. (1997). Controlled unsteady state processes and technologies - an overview. *Wat. Sci. Tech.*, **35**, 1-10.

Kong Y.H., Beer M., Seviour R.J., Lindrea K.C., and Rees G.A. (2002). Role of "G-bacteria" in anaerobic substrate uptake in a SBR with no phosphorus removal. *Wat. Sci. Tech.*, **46** (1-2), 171-178.

Liu W.-T., Mino T., Nakamura K., and Matsuo T., and (1996). Glycogen accumulating population and its anaerobic substrate uptake in anaerobic-aerobic activated sludge without biological phosphorus removal. *Wat. Res.*, **30** (1), 75-82

Liu W.-T., Nakamura K., Matsuo T., and Mino T. (1997). Internal energy-based competition between bolyphosphate- and glycogen-accumulating bacteria in biological phosphorus removal reactors---effect of P/C feeding ratio. *Wat. Res.*, **31** (6), 1430-1438.

Seviour R.J., Mino T., Onuki M. (2003). The microbiology of biological phosphorus removal in activated sludge systems. *FEMS Microbiology Reviews*, **27**, 99-127.

Standard Methods for the Examination of Water and Wastewater (1998). 20th ed., American Public Health Association/American Water Works Association/Water Environment Federation, Washington DC, USA.

Stephens H. L. and Stensel H.D. (1998). Effect of operating conditions n biological phosphorus removal. *Water Environment Research*, **70** (3), 362-369.

Sudiana I. M., Mino T., Satoh H., Nakamura K., and Matsuo T. (1999). Metabolism of enhanced biological phosphorus removal and non-enhanced biological phosphorus removal sludge with acetate and glucose as carbon source. *Wat. Sci. Tech.*, **39** (6), 29-35.

Wilderer P.A., Irvine R.L., and Goronszy M.C. (2001). Sequencing batch reactor technology. Scientific and technical report No.10, IWA publishing, London, England.

Zhu J. and Liu C. (1999). Cultivation and physic-chemical characteristics of granular activated sludge in alternation of anaerobic/aerobic process. *Chinese Environmental Science*, **20** (3), 38-41.

Zhu J. and Wilderer P.A. (2003). Effect of extended idle conditions on structure and activity of granular activated sludge. *Wat. Res.*, **37**, 2013-2018.

Developing Aerobic Phenol-Degrading Granules from Acetate-Fed Granules

Stephen Tiong-Lee Tay, Benjamin Yan-Pui Moy, Abdul Majid Maszenan and Joo-Hwa Tay

Environmental Engineering Research Centre, School of Civil and Environmental Engineering, Nanyang Technological University, 50 Nanyang Avenue, Singapore 639798 (e-mail: ctltay@ntu.edu.sg)

Abstract Aerobic granules cultivated previously on acetate were used as a microbial inoculum in a sequencing batch reactor to develop aerobic phenol-degrading granules. The reactor was operated using 4 hr cycles with phenol as sole carbon and energy source. The phenol loading rate was 1.8 kg m^{-3} d^{-1} and the influent phenol concentration was 600 mg L^{-1}. There was a lag of two days when the acetate-fed granules failed to degrade phenol, but the biomass acclimated quickly and completely degraded phenol thereafter. Stable phenol-degrading granules were obtained two weeks after start-up. The acetate-fed granules were initially dominated by bacterial rods, but exposure to high concentrations of toxic phenol triggered the proliferation of filamentous bacteria. Denaturing gradient gel electrophoresis of PCR-amplified 16S rRNA gene fragments showed that the acetate-fed granules and the phenol-degrading granules had different microbial communities.

Keywords Aerobic Granules; Microbial Inoculum; Phenol Degradation; Substrate Inhibition

Introduction

Phenol is widely used in the production of polycarbonate resins, explosives, paints, inks, perfumes, wood preservatives, textiles and drugs, and as an antibacterial and antifungal agent or disinfectant. Industries that use phenol may at some point release this compound into the environment. For instance, phenol concentrations of up to 10,000 mg L^{-1} have been reported in industrial wastewater effluents (Fedorak and Hrudey, 1988). Without proper treatment, industrial wastewaters would become an important source of anthropogenic phenol into the environment. Phenol-containing wastewater is difficult to treat biologically because of substrate toxicity and inhibition. However, these substrate inhibition difficulties can be overcome by strategies such as cell immobilization. Aerobic granulation represents a relatively recent form of cell immobilization that is attracting considerable research attention (Beun *et al.*, 1999; Moy *et al.*, 2002).

Aerobically grown microbial granules are self-immobilized aggregates of bacteria cultivated in sequencing batch reactors (SBRs) and have a strong, compact microbial structure, good settling ability and high biomass retention, with the ability to handle high organic loading rates. Jiang *et al.* (2002) first documented the successful cultivation of aerobic granules with toxic phenol as a sole carbon source at a relatively modest loading rate of 1.5 kg phenol m^{-3} d^{-1}. These phenol-degrading aerobic granules possessed good tolerance against high phenol concentrations, and the kinetics data indicated that these granules had the potential to treat industrial wastewaters with high phenol loads. However, these phenol-degrading granules required a total start-up of three months, with two months for conditioning of municipal activated sludge seed by incubating with phenol, and one month for in-reactor development of stable granules from the conditioned sludge seed.

Such long start-up times may pose a problem in deploying aerobic granulation for industrial application. Not much is known about how start-up times are controlled in aerobic granulation systems, although start-up times in anaerobic granulation systems are known to depend on the environmental and hydrodynamic conditions inside the reactor, including the type of wastewater, the liquid upflow velocity, the hydraulic retention time, the amount of seed sludge and the use of suitably acclimated seed as microbial inocula (Cronin and Lo, 1998; Ghangrekar *et al.*, 1996). The main objective of the current study was therefore to assess the suitability of using acetate-fed granules as a microbial inoculum for the development of aerobic phenol-degrading granules. This work should be of practical interest and could contribute to a better understanding of how aerobic granulation technology can be exploited in the field.

Materials and Methods

Reactor operation

Experiments were performed in a column-type sequencing batch reactor (SBR) at 25°C. The reactor was 150 cm tall with an internal diameter of 5 cm and a working volume of 2.5 L. The reactor was seeded with 0.5 L of acetate-fed aerobic granules cultivated as described previously (Moy *et al.* 2002). The initial biomass concentration was approximately 4 g suspended solids (SS) L^{-1}. Fine air bubbles for aeration and mixing were supplied through a dispenser at the reactor bottom at an airflow rate of 3.5 L min^{-1}, equivalent to a superficial gas velocity of 3 cm s^{-1}. The reactor was operated sequentially in 4 hr cycles, with 4 min of influent filling, 227 min of aeration, 5 min of settling and 4 min of effluent withdrawal. Effluent was discharged 60 cm above the reactor bottom at a volumetric exchange ratio of 50%, giving a hydraulic retention time of 8 hr. The reactor was operated at a phenol loading rate of 1.8 kg phenol m^{-3} d^{-1}, equivalent to a COD loading rate of 4.28 kg COD m^{-3} d^{-1} (COD stands for chemical oxygen demand). The synthetic wastewater had the following composition: phenol, 600 mg L^{-1}; NH_4Cl, 100 mg L^{-1}; K_2HPO_4, 22.5 mg L^{-1}; $CaCl_2.H_2O$, 15 mg L^{-1}; $MgSO_4.7H_2O$, 12.5 mg L^{-1}; $FeSO_4.7H_2O$, 10 mg L^{-1}; and 1 ml L^{-1} micro-nutrients (Moy *et al.* 2002).

Analytical methods

Wastewater samples were analyzed for suspended solids (SS), volatile suspended solids (VSS) and sludge volume index (SVI) using Standard Methods (APHA, 1998). Specific oxygen utilization rate (SOUR) was determined by incubating batch samples with 300 mg phenol L^{-1} in standard BOD (biochemical oxygen demand) bottles. Dissolved oxygen (DO) levels were measured with a DO meter (YSI 5000; YSI Incorporated, Yellow Springs, OH, USA) at time intervals of 10 s over a 10 min period. SOUR was calculated by dividing the oxygen consumption rate by the VSS concentration (Jiang *et al.*, 2002). Phenol was measured spectrophotometrically by the 4-aminoantipyrine method (APHA, 1998). Samples were centrifuged at 4°C for 5 min (14,000 *g*) and 5 ml of supernatant withdrawn. 100 μL of 0.5N NH_4OH was added to the supernatant and the pH was adjusted to 7.9±0.2 with phosphate buffer. 0.5 mL of 0.4% (w/v) 4-aminoantipyrine and 0.5 mL of 1.6% (w/v) $K_3Fe(CN)_6$ were then added. The contents were well mixed and allowed to stand for 15 minutes before measuring the absorbance at 500 nm. Phenol concentrations were then determined by reference to standard curves. Granule microstructure was observed with scanning electron microscopy (Stereoscan 420, Leica Cambridge Instruments). Granules were prepared by washing with a phosphate buffer and fixing with 2% glutaraldehyde overnight at 4°C. Fixed granules were washed with 0.10 M sodium cacodylate buffer, dehydrated by successive passages through 25, 50, 75, 80, 90, 95 and 100% ethanol and dried with a critical point dryer (Polaron E3000, VG Microtech).

DNA extraction and denaturing gradient gel electrophoresis (DGGE)

Genomic DNA of granule biomass was extracted by bead beating method as described previously (Tay *et al.*, 2002). PCR primers P2 and P3 (containing 40 bp of GC clamp) (Muyzer *et al.*, 1993) were used to amplify the variable V3 region of bacterial 16S rDNA (corresponding to positions 341 to 534 in the *Escherichia coli* sequence). PCR conditions and thermal programs for DGGE have been previously described (Watanabe *et al.*, 1998). The DGGE profiles were analyzed using GelCompar software (Applied Maths BVBA, Sint-Martens-Latem, Belgium).

Results

Reactor performance

The SBR was operated with phenol as sole carbon and energy source. A constant loading rate of 1.8 kg phenol m^{-3} day^{-1} was maintained, corresponding to an influent phenol concentration of 600 mg L^{-1}. The reactor showed good biomass retention, and the biomass concentration stabilized at 4.6 g L^{-1} two weeks after start-up (Table 1; Fig. 1). The reactor experienced some initial difficulty in removing phenol, as both start-of-cycle and end-of-cycle phenol concentrations increased from 300 mg L^{-1} to 500 mg L^{-1} in the first two days of reactor operation (Fig. 2).

However, the start-of-cycle and end-of-cycle phenol concentrations rapidly declined to stabilize at 326 mg L^{-1} and 0.3 mg L^{-1}, respectively, from day 3 of reactor operation. Fig. 3 shows the profile of phenol concentration within a single 4 hr cycle on day 30. Phenol was rapidly removed in the reactor, and the phenol concentration decreased from 313 mg L^{-1} initially to less than 0.4 mg L^{-1} after 45 min into the cycle.

Table 1: Reactor performance and granule characteristics at steady state

Reactor and granule parameters	Values
Biomass concentration (g SS L^{-1})	4.62±0.59
Start-of-cycle phenol concentration (mg phenol L^{-1})	326±19
End-of-cycle phenol concentration (mg phenol L^{-1})	0.3±0.1
Phenol removal efficiency (%)	99.9±0.1
Specific oxygen utilization rate (mg O_2 g^{-1} VSS h^{-1})	54±6
Sludge volume index (mL g^{-1})	70±6

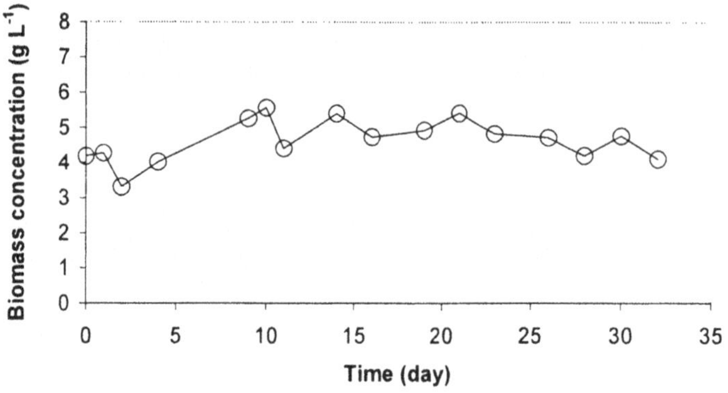

Figure 1: Biomass profiles

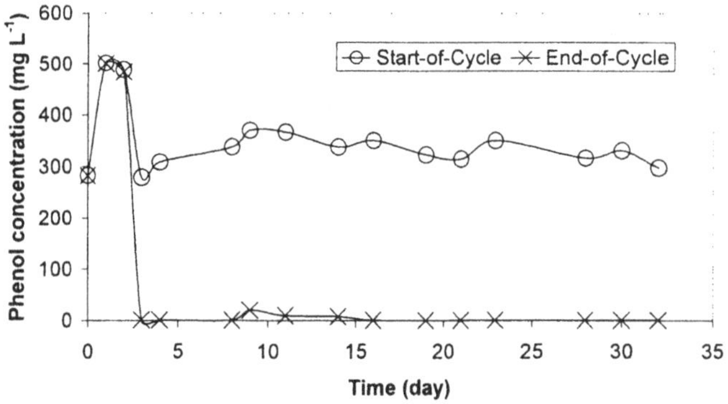

Figure 2: Phenol concentration profiles

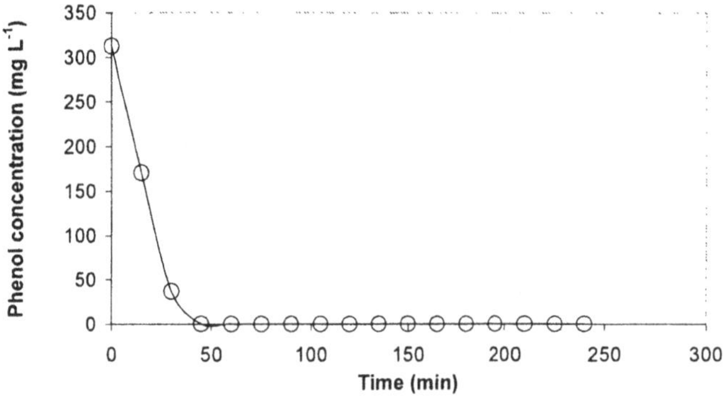

Figure 3: Phenol removal in one cycle on day 30

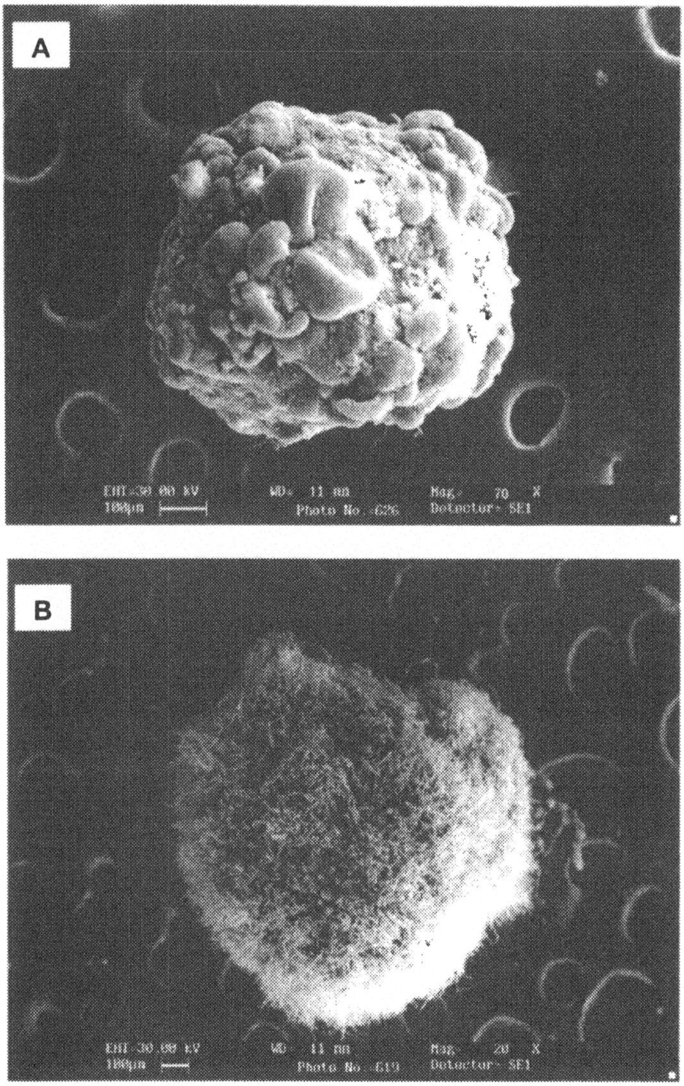

Figure 4: Scanning electron micrographs of aerobic granules on (A) day 0; (B) day 30

Development of phenol-degrading granules

The acetate-fed granules that were used to seed the reactor displayed a lumpy appearance (Fig. 4A) and seemed to consist of small colonies agglomerated together. The bacterial community on the granule surface was initially dominated by bacterial rods (Fig. 5A), which was later replaced by a thick envelope of filamentous bacteria (Fig. 4B and Fig. 5B). SOUR and SVI averaged 54 mg O_2 g VSS^{-1} h^{-1} and 70 mL g^{-1}, respectively (Table 1).

DGGE profiles

The DGGE results provide information on the composition of the dominant members in the bacterial community (Fig. 6). There was an obvious change in the community composition as only 2 comigrating bands were found that were common to granules sampled at the beginning and the end of the reactor operation.

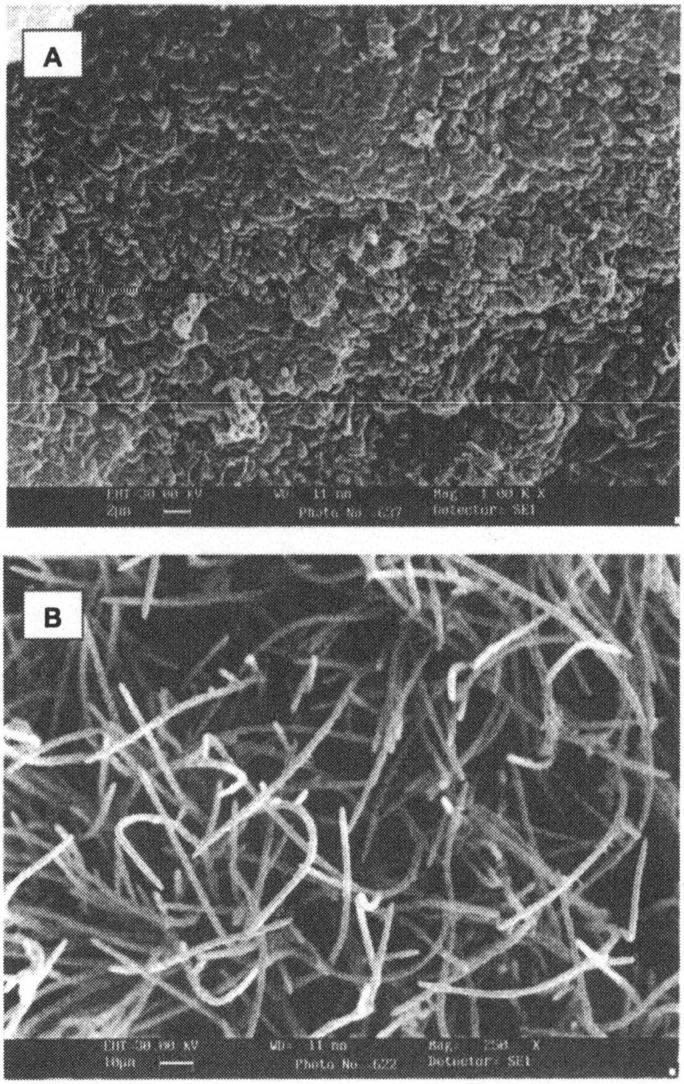

Figure 5: Scanning electron micrographs of granule surfaces on (A) day 0; and (B) day 30

Discussion

In this study, we explored the use of aerobic acetate-fed granules as an inoculum for the cultivation of aerobic phenol-degrading granules. This approach produced stable phenol-degrading granules with good settling ability, good biomass retention and good metabolic activity, as evidenced by the low SVI values, stable biomass concentrations and good phenol removal. The strong and compact structure of the acetate-fed granules likely provided the microorganisms with adequate protection against phenol toxicity. Within these compact granules, a phenol concentration gradient is developed because of diffusional resistance, and this serves to protect the microorganisms by diluting the toxic chemical below some threshold value to avoid inhibition and allow continued microbial activity and substrate utilization (Jiang *et al.*, 2002). Although there was a slight lag in the ability of the acetate-fed granules to degrade phenol initially, presumably to allow the buildup of a critical population of phenol-degrading microorganisms, the granules quickly acclimated to the phenol load and degraded phenol completely three days after start-up. The granules stabilized within two weeks after start-up, with little change in biomass concentration, phenol removal and specific mineralization activity.

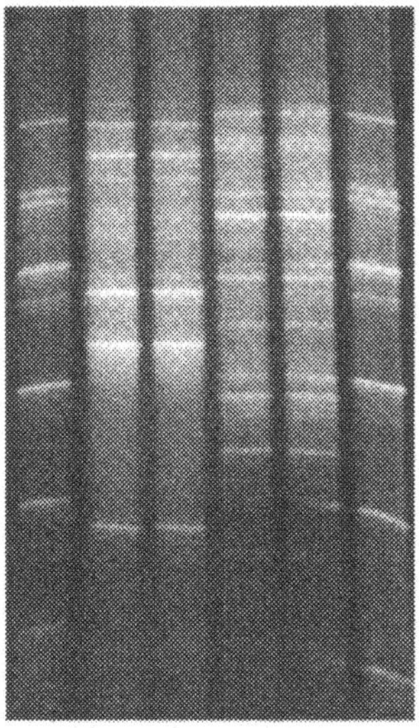

Figure 6: DGGE profiles of partial 16S rRNA gene fragments. Lanes: 1, migration standards; 2 and 3, day 0 granules; 4 and 5, day 30 granules; 6, migration standards

The granules adapted quickly to survive and thrive with phenol as sole substrate. The ability of the granules to recover from phenol toxicity showed that the toxicity effects exerted by the high phenol concentrations were neither permanent nor irreversible. Moreover, the sudden exposure to high phenol concentrations did not result in a reduction in species richness in the granules, as measured by the number of DNA bands detected by DGGE. Since functional redundancy can be defined as the number of different species that perform a specified function, the richness of bacteria residing in the phenol-degrading granules can be used as a measure of this parameter (Yin *et al.*, 2000). The community structure banding patterns from the DGGE experiments showed that the phenol-degrading granules that developed from acetate-fed granules also contained a high variety of bacteria that responded favorably to high concentrations of phenol. Indeed, biodiversity exerts a profound influence on ecosystem stability and productivity, and a more diverse community should be more resistant to external stresses (Tilman and Downing, 1994). Our results point to the versatility of using acetate-fed granules as a starting seed for developing granules for the biodegradation of other toxic chemicals.

Filamentous bacteria are slow-growing microorganisms whose growth is generally considered to be favored by low nutrient or low oxygen conditions (Jenkins *et al.*, 1993; Martins *et al.*, 2004). In systems where the substrate concentration is high, like in plug-flow reactors and the SBR system used in the current study, the filamentous bacteria should be suppressed since their growth rate is expected to be lower than that for the floc-forming bacteria. Therefore, considering the high concentrations of phenol substrate prevailing in the SBR used in this study, the emergence and eventual dominance of filamentous bacteria in the granules was unexpected. The precise reasons for the dominance of the filamentous bacteria in the phenol-degrading granules will have to be resolved in a future investigation.

In summary, acetate-fed granules served as a good starting inoculum for the development of granules targeted for the biodegradation of high concentrations of toxic phenol. This strategy of using granules fed on benign acetate to produce granules that can degrade toxic phenol can be extended to granule-based applications that target other toxic chemicals.

Acknowledgements

This work was supported by Nanyang Technological University funds to S.T.L. Tay and J.H. Tay. We thank Eileen Hui-Ling Teng for assisting with the analytical protocols.

References

APHA (1998). *Standard methods for the examination of water and wastewater*, 20th edn. American Public Health Association, Washington, DC.

Beun J.J., Hendriks A., van Loosdrecht M.C.M., Morgenroth E., Wilderer P.A. and Heijnen, J.J. (1999). Aerobic granulation in a sequencing batch reactor. *Water Res.*, **33**, 2283-2290.

Cronin, C. and Lo, K.V. (1998). Anaerobic treatment of brewery wastewater using UASB reactors seeded with activated sludge. *Bioresource Technol.*, **64**(1), 33-38.

Fedorak, P.M. and Hrudey, S.E. (1988). Anaerobic degradation of phenolic compounds with application to treatment of industrial waste waters. In: *Biotreatment Systems, Vol. 1.*, D.L. Wise (ed.), CRC Press, Boca Raton, FL., pp. 170-212.

Ghangrekar, M.M., Asolekar, S.R., Ranganathan, K.R. and Joshi, S.G. (1996). Experience with UASB reactor start-up under different operating conditions. *Water Sci. Technol.*, **34**(5-6), 421-428.

Jenkins D., Richard M.G. and Daigger G.T. (1993). *Manual on the Causes and Control of Activated Sludge Bulking and Foaming.* (2nd edn.). Lewis Publ. Boca Raton, FL, USA.

Jiang H.L., Tay J.H. and Tay S.T.L. (2002). Aggregation of immobilized activated sludge cells into aerobically grown microbial granules for the aerobic biodegradation of phenol. *Lett. Appl. Microbiol.*, **35,** 439-445.

Martins A.M.P., Pagilla K., Heijnen J.J. and Van Loosdrecht M.C.M. (2004). Filamentous bulking sludge – a critical review. *Water Res.*, **38**, 793-817.

Moy B.Y.P., Tay J.H., Toh S.K., Liu Y. and Tay S.T.L. (2002). High organic loading influences the physical characteristics of aerobic sludge granules. *Lett. Appl. Microbiol.*, **34**, 407-412.

Muyzer G., de Waal E.C. and Uitterlinden A.G. (1993). Profiling of complex microbial populations by denaturing gradient gel electrophoresis analysis of polymerase chain reaction-amplified genes coding for 16S rRNA. *Appl. Environ. Microbiol.*, **59,** 695-700.

Tay S.T.L., Ivanov V., Yi S., Zhuang W.Q. and Tay J.H. (2002). Presence of anaerobic Bacteroides in aerobically grown microbial granules. *Microb. Ecol.*, **44**, 278-285.

Tilman D. and Downing J.A. (1994). Biodiversity and stability in grasslands. Nature, **367**, 363-365.

Watanabe K., Teramoto M., Futamata H. and Harayama S. (1998). Molecular detection, isolation, and physiological characterization of functionally dominant phenol-degrading bacteria in activated sludge. *Appl. Environ. Microbiol.*, **64**, 4396-4402

Yin B., Crowley D., Sparovek G., De Melo W.J. and Borneman J. (2000) Bacterial functional redundancy along a soil reclamation gradient. *Appl. Environ. Microbiol.*, **66,** 4361-4365.

Treatment of food industry effluents in a granular sludge SBR

N Schwarzenbeck* and PA Wilderer*

*Institute of Water Quality Control and Waste Management, Technical University of Munich, Am Coulombwall, 85748 Garching, Germany (E-mail: N.Schwarzenbeck@bv.tum.de)

Abstract Aerobic granular sludge is a promising technology to overcome limitations that are inherent to the nature of the activated sludge process. High settling velocities of the microbial aggregates allow the operation of activated sludge reactors at higher MLSS-contents and simultaneously at more compact sedimentation processes. At the same time slow growing, but specialized microorganisms can be immobilized within the biofilm matrix of the sludge granules. This makes the technology particularly interesting for the treatment of industrial wastewaters. Within this study, the applicability of aerobic granular sludge to the treatment of food industry effluents was investigated. Experiments with malting wastewater showed the development of a stable granular sludge bed that could remove both dissolved and particulate substrates sufficiently. Tracer experiments revealed a dense sessile protozoa population covering the granules, which appeared to be responsible for primary particle uptake from the wastewater. From experiment with dairy wastewater a coherence between substrate degradation kinetics and granule structure could be derived. As a result the maximum applicable organic loading rate is limited as slow substrate uptake kinetics result in filamentous outgrowth and granule instability. As for SBR-systems volumetric exchange ratio and organic loading rate are coupled, the process efficiency is limited. Attention has also to be paid to the fact, that suspended solids are always present in the effluent of aerobic granular sludge reactors, making a post treatment step necessary.

Keywords Aerobic granular sludge, Food industry effluents, SBR

Introduction

The activated sludge process is probably the most commonly applied and best investigated process in biological wastewater treatment. It is the process of choice for the treatment of municipal wastewaters as well as for industrial effluents, where anaerobic and physico-chemical processes are economically not favorable (Rosenwinkel 2000). Although the activated sludge process has been subject to intensive research over the past decades, there are some principal limitations inherent to the process. The performance of the process is significantly dependent upon two factors: The metabolic capabilities of the microbial community in the system and the efficiency of the solid-liquid separation (biomass retention) at the last stage of the treatment procedure. For the latter the gravity-driven sedimentation process is commonly applied. In this case the efficiency of biomass separation is a function of the settling characteristics of the microbial aggregates. Biomass retention on the other hand can be used for improving the removal capacities of a system by increasing the overall biomass content, and by fostering the accumulation of specialized but slow growing organisms. One approach to enhance the performance of both factors at the same time is the cultivation of aerobic granular sludge in activated sludge reactors. As granular sludge settles impressively faster than activated sludge flocs, a higher biomass retention can be achieved within a more condensed settling process. Furthermore are microorganisms immobilized in the biofilm matrix of the granular sludge aggregates, which allows the hold-back of slow-growing but specialized organisms in the reactor. Whereas in conventional activated sludge process an increase in biomass content (e.g. using membrane filtration) is accompanied by an increase in viscosity and a decrease in oxygen transfer coefficient, this effect could not be observed for aerobic granular sludge. Of additional interest is the ability of aerobic granules to keep metabolic activity over an anaerobic storage period of several months (Zhu et al 2003). Particularly industries with seasonal production (e.g. starch producing industries) are possible areas of application. Granular sludge can hence be concluded an appropriate measure for realizing high rate aerobic treatment systems and for enhancing the capacity of an existing system that is limited by its volumetric conversion capacities.

Diverse research has been conducted so far on the formation of aerobic granular sludge under various synthetic conditions (Shin et al 1992, Morgenroth et al 1997, Peng et al 1999, Etterer et al 2001, Tay et al 2001, Beun et al 2002, Tsuneda et al 2003, Zhu et al 2003, de Kreuk et al 2003, Yang et al 2004). The formation of microbial aggregates is induced by a number of stressors like high DO-concentrations, a considerable level of hydrodynamic shear stress and a frequent repetition of distinct feast and famine conditions as associated with the intermittent feeding of the SBR-cycle (Bossier et al 1996). Once granules have started to form, they can specifically be selected for, by applying a threshold settling velocity, below which all slower settling biomass is washed out from the system (Beun et al 2002, Tay et al 2002). Overall all commonly applied processes of aerobic wastewater treatment (COD-removal, nitrification/denitrification and enhanced biological phosphorus removal EBPR) have been observed for aerobic granular sludge reactors under synthetic conditions. Knowledge on the behaviour of aerobic granules in treatment systems for real wastewaters is nevertheless scarce up to now. Little attention has been paid to the fate and the influence of particulate and colloidal organic matter. This aspect is on the other hand crucial, when treating real wastewaters, where major fractions are present in a non-dissolved form. Little attention has also been paid to facets of large scale application. Amongst these are aspects such as reactor behaviour during start-up, MLSS-content in the effluent after complete granulation (slow settling biomass has to be washed out from the system continuously) and the formation of distinct feast and famine conditions as a function of loading rate, which again for SBR-systems is dependent on the volumetric exchange ratio.

Material and Methods

Experimental setup

Malting wastewater.

Experiments with malting wastewater were conducted in a lab-scale SBR with a total working volume of 12 L at an internal diameter of 20 cm and a filling height of 38 cm. The reactor was aerated through air bubble diffusers at a volumetric flow rate of 600-800 L/h (superficial upflow air velocity 4.4-5.9 cm/s). Total cycle duration was 8 h, with 6 min of fill, 120 min of anaerobic no mix, 345-348 min of aeration, 5-2 min of settling and 4 min of effluent withdrawal. Volumetric exchange ratio (VER) was set to 66 %. Average COD_{total} loading rate was 3.2 kg/(m3·d), wastewater was fed from the top of the reactor. Wastewater was prepared as a concentrate in a storage tank (V=50 L) and during fill period diluted with tap water 1:2.

Dairy wastewater.

Experiments with dairy wastewater were carried out in a bubble column reactor (76.0 cm filling height and 14.2 cm diameter) with a working volume of 12.0 L and volumetric exchange ratios of 75 % to 50 %. Wastewater was fed from a storage tank (50 L volume) and introduced at the bottom of the reactor. Aeration was provided by means of air bubble diffusers at a superficial upflow air velocity of 1.1 cm/s. The reactor was operated in sequencing batch mode with a total cycle duration of 8 hours at 60 min fill, 60 min anaerobic no-mix, 335-405 min of aeration, 15-4 min settling, 5 min decant and 5 min of idle. The volumetric exchange ration was between 25 % and 75 %. Average COD_{total} loading rate was between 5.8 kg/(m^3·d) and 2.9 kg/(m^3·d).

Both experiments.

Temperature in the reactors was ambient (17 °C ± 3 °C). Activated sludge (V = 12 L) from a municipal wastewater treatment plant was used as inoculum. Biofilm growing on the reactor walls was removed from the reactor every fortnight while storing the reactor content separately.

Wastewater

The wastewater used in the malting wastewater experiment was prepared as a concentrate by mixing barley dust together with tap water. Dairy wastewater was collected from an industrial dairy plant once a week and stored at +2 °C to +6 C. Prior to feeding into the reactor pH was adjusted to around 8.0 (± 1). Table 1 shows the average composition of the wastewaters used throughout the experiment in comparison with values from the literature (Rosenwinkel 2000).

Table 1: Average composition of the wastewaters in this study compared to the literature

		COD_{total}	$COD_{dissolved}$	BOD_5	N_{total}	P_{total}	MLSS
		mg/L	mg/L	mg/L	mg/L	mg/L	mg/L
Dairy	this study	2800	1500	1600	140	30	300
	literature	500 – 4500	n.s.	500 – 2000	30 – 250	10 – 100	n.s.
Malting	this study	1700	470	700	50	9	950
	literature	925-1732	n.s.	600-800	n.s.	n.s.	1258

n.s. = not specified

Analytical methods

Total nitrogen (N_{total}), total phosphorus (P_{total}) and chemical oxygen demand (COD) were measured spectrophotometrically (Dr. Lange test equipment and Dr. Lange photometer ISIS 6000). For analysis of dissolved parameters samples were filtered at 0.45 μm pore size prior to analysis. In order to obtain an appropriate measure for the removal of total parameters from the dairy wastewater, the effluent was also analyzed after two hours of settling in an Imhoff-cone. Mixed liquor suspended solids (MLSS) and volatile suspended solids (VSS) were measured according to German standard methods (DEV). Sludge settling characteristics were determined according to a modified procedure of the sludge volume index (SVI) measurement, the dynamic SVI: Sludge was filled into a graduated cylinder (V=1 L), sludge volume was recorded every 5 minutes for a duration of 30 minutes. Values of sludge volume were subsequently normalized by the MLSS content of the sample. The influence of sedimentation time on the effluent quality was determined in an Imhoff-cone. For detailed investigations of the COD-removal of particulate substrates, COD-fractions were separated by means of particle size and settleability according to Roessink et al. (1997).

Results and Discussion

Granule formation and granule structure

For both experiments smooth and dense granules appeared within the first week of operation, whereas for both wastewaters it took several weeks to develop a completely granular sludge bed. Granules grown on particulate rich malting wastewater showed a high number and diverse population of, mainly sessile, protozoa colonizing the granules surface. Whereas for dairy wastewater containing less suspended solids only a minor protozoa population could be found. Similar observations were made by other researchers for particulate biofilm reactors (Roessink et al 1997) fed with wastewaters containing different levels of particulate matter. For granules grown on dairy wastewater on the other hand strong tendencies towards the development of outgrowing filamentous structures could be observed.

Sludge settling and structural characteristics of the granulizing sludge bed

With regards to the design of a large scale wastewater treatment unit that is operated with aerobic granular sludge, an adequate method has to be found, that reliably and objectively describes the settling characteristics of the sludge bed. Ideally the same method can also be used to monitor the structural characteristics of the sludge bed (granules vs. flocs) during operation and indicates changes that might lead to an operational failure. So far only static parameters such as SVI and settling velocities of single granules have been incorporated for this purpose. Meanwhile both methods do not include the dynamic interactions between the moving liquid and between individual sludge aggregates, which finally determine the settling rate of an activated sludge bed. Throughout this study the change of SVI over a settling period of 30 minutes has shown to be an adequate parameter to implement all the mentioned goals in one tool. Figure 1 shows exemplary the development of the SVI at different settling times along the duration of the experiment with dairy wastewater together with the respective structural sludge characteristics.

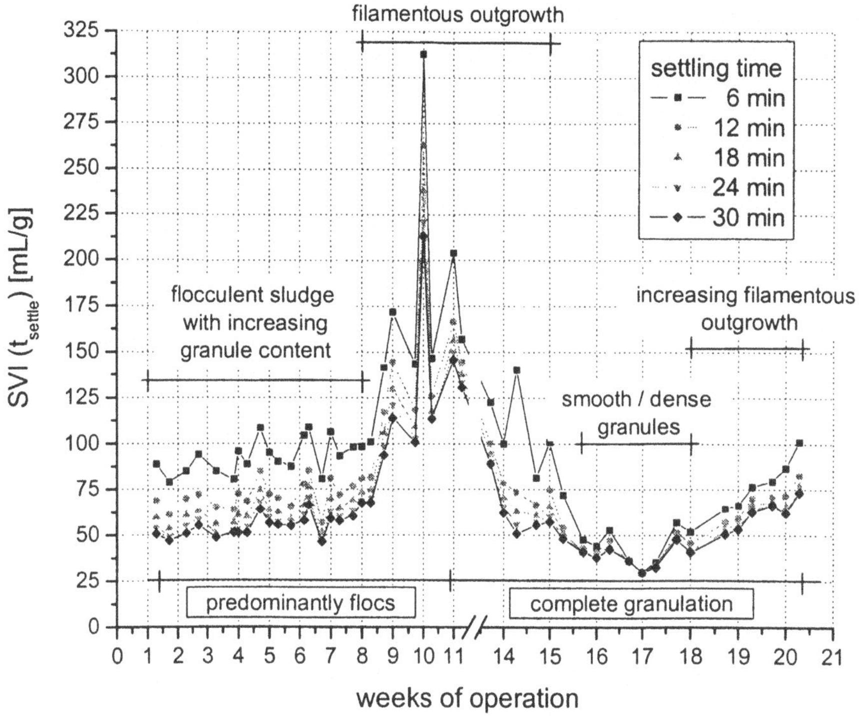

Figure 1: Aerobic granular sludge SBR - relation between sludge settling characteristics and structural properties

Reactor performance

COD-removal after development of a completely granular sludge bed.

In both experiments comparably high removal efficiencies could be achieved, taking into account the organic loading rates that were applied and are considerably higher than those used for the design of real scale activated sludge systems treating the same types of wastewater. In the malting wastewater experiment average removal efficiencies of 50 % in COD_{total} and 80 % in $COD_{dissolved}$ were achieved at a biomass concentration (MLSS) of 6-7 g/L, an organic loading rate (COD_{total}) of 3.4 kg/($m^3 \cdot d$) and an influent particle concentration (TSS) of 0.9 g/L. In the dairy wastewater experiment average removal efficiencies of 67 % in COD_{total} and 93 % in $COD_{dissolved}$ were achieved at a biomass concentration (MLSS) of 7-8 g/L and an organic loading rate (COD_{total}) of 5.8 kg/($m^3 \cdot d$). Table 2 gives an overview of the effluent qualities.

Table 2: Effluent quality of the granular sludge SBRs after complete granulation

	COD_{total}	$COD_{total,\ 2\ h\ settling}$	$COD_{dissolved}$	MLSS
	mg/L	mg/L	mg/L	mg/L
malting	950	75	65	450
dairy	1100	217	100	640

In order to maintain a completely granular sludge bed, very particular operating conditions have to be applied, i.e. slow settling biomass has to be continuously washed out from the system. As a result the effluent quality immediately after discharge from the reactor is very much determined by the biomass content in the effluent, which again is dependent on the settling characteristics of the sludge aggregates. As a result the effluent quality immediately after discharge is not satisfactory and a post treatment step is required. The relation between MLSS content and COD_{total} in the effluent and the influence of separation of the biomass from the effluent can be seen from table 2 and table 3 exemplarily for COD_{total}, N_{total} and P_{total}.

Nutrient removal.

The raw malting wastewater shows only minor concentrations of nutrients at a ratio of C:N:P = 100:3:0.5. The removal of nutrients can hence be attributed to incorporation during the growth of new biomass. As the raw dairy wastewater on the other hand contains significant concentrations of nutrients, the removal efficiency in this respect is also of great interest. For determining the removal efficiency, the same approach as for COD_{total} was applied (sedimentation of the settleable solid matter). Table 3 shows the concentration of N_{total} and P_{total} in the influent and effluent of the experimental system. Again it can be seen, that the overall removal efficiency is very much dependent on the removal of solid matter from the effluent.

Table 3: Nutrient removal in the SBR treating dairy wastewater after complete granulation

	N_{total} mg/L	$N_{total,\ 2h\ settling}$ mg/L	P_{total} mg/L	$P_{total,\ 2h\ settling}$ mg/L
influent	130	130	23	23
effluent	110	50	15	5

Removal of particulate matter.

Real wastewaters contain significant parts of the pollution in form of particulate or colloidal matter. The removal mechanism for particulate matter in granular sludge SBRs was exemplarily studied using malting wastewater. It could be shown, that removal efficiency is a function of particle size and particle degradability. The details are shown in table 4.

Table 4: COD-removal efficiency in the granular sludge SBR treating malting wastewater as a function of particle size

	total	D_P>25-50μm	D_P<25-50μm	1μm>D_P>0.2μm
$COD_{influent}$ [mg/L]	1700	1550	65	120
$COD_{effluent}$ [mg/L]	950	875	18	2

The ability of aerobic granules to remove particulate matter from the wastewater could be attributed to two different mechanisms: During biofilm formation and growth particles are incorporated into the biofilm matrix. After complete granulation has occurred a diverse population of mostly sessile protozoa could be observed, whose metabolic activity revealed to be responsible for the removal of particulate matter from the wastewater.

Organic loading rate.

Previous research indicates that smooth and dense granules can be obtained for organic loading rates of up to 15 $kgCOD_{total}/(m^3 \cdot d)$ (Moy et al 2002). Within the present study filamentous granules formed in dairy wastewater at much lower average loading rates of 5.8 $kgCOD_{total}/(m^3 \cdot d)$. Filamentous organisms show a higher maximum growth rate at lower substrate concentrations than floc-forming organisms, whereas floc-forming organisms show higher growth rates at higher substrate concentrations. As a result the growth of filamentous organisms is fostered when intermediate substrate concentrations are available throughout long parts of the SBR-cycle. Dairy wastewater unlike synthetic wastewater shows a gentle slope of the substrate gradient between feast and famine conditions throughout the SBR-cycle. This can be attributed to the fact, that real wastewaters show a rather different behaviour in biological degradability from the one of synthetic wastewaters. In the first case significant parts of the organic pollution are present in form of colloids and particles and have to be slowly hydrolyzed prior to cell-internal metabolism, whereas in the second case only dissolved and readily degradable substrates are used

Effluent quality and post treatment.

The SBR-process allows a very compact design of wastewater treatment units, implicating the actual biological reactor and the solid/liquid separation process within one single constructional element (Wilderer et al 2001). SBR-systems for the formation of aerobic granules have on the other hand to be operated in a way, which washes out slow settling biomass (i.e. sludge flocs) every cycle and thus discharges the excess sludge together with the effluent. As a result a significant fraction of suspended solids can be found in the effluent, resulting in comparably high COD_{total} and N_{total} values (see also tables 2 and 3). An immediate discharge of effluents from aerobic granular sludge reactors directly into receiving waters, as proposed in the past, is hence impossible, sufficient post-treatment has to be provided. During this investigations secondary clarification with a hydraulic retention time between 15 and 30 minutes was found to be sufficient to separate the biomass from the treated wastewater and to reach the maximum achievable effluent quality. Applying pretreatment to the wastewater can only help improving the effluent quality if the raw wastewater contains significant parts of settleable matter.

Conclusions

- Aerobic granular sludge can successfully be cultivated in SBR-systems treating real wastewaters containing particulate and colloidal organic matter.
- The structural characteristics of a granulising sludge bed as well as the structural characteristics of the individual sludge aggregates can be appropriately be described in terms of a modified procedure of SVI-measurement (i.e. change in SVI over a settling period of 30 minutes) rather than SVI itself.
- A distinct and steep substrate gradient between periods of low and high substrate availability ("feast and famine") is the key factor inducing the formation of smooth and dense granules.

- Post-treatment is required for aerobic granular sludge reactors as the effluent contains significant concentrations of biomass. Nevertheless a sedimentation process with 15 to 30 minutes HRT is sufficient to achieve maximum possible effluent quality.
- Protozoa play an important role in the removal and metabolic conversion of particulate matter from wastewaters in granular sludge bed reactors as they are the venue of primary particle uptake.
- Process efficiency in terms of volumetric exchange ratio and cycle duration is limited in terms of feast and famine conditions as granules show structural deficiencies at high organic loading rates due to filamentous outgrowth.

Acknowledgements

This work was financially supported by the German research foundation (Deutsche Forschungsgemeinschaft DFG), grants Wi 620/15-1 and 2

References

Beun J.J., van Loosdrecht M.C.M., Heijnen J.J. (2002). Aerobic granulation in a sequencing batch airlift reactor. *Wat. Res.* **36**(3), 702-12

Bossier P, Verstraete W (1996). Triggers for microbial aggregation in activated sludge?. *Appl. Microbiol. Biotechnol.* **45**(1-2), 1-6

De Kreuk M.K., van Loosdrecht M.C.M. (2004). Selection of slow growing organisms as a means for improving aerobic granular sludge stability. *Wat. Sci. Tech.* **49**(11-12), 9-17

Deutsche Einheitsverfahren zur Wasser-, Abwasser- und Schlammuntersuchung (German Standard Methods for the Examination of Water and Wastewater). Fachgruppe Wasserchemie in der Gesellschaft Deutscher Chemiker, VCH Verlag Chemie, Weinheim.

Etterer T., Wilderer P.A. (2001). Generation and properties of aerobic granular sludge. *Wat. Sci. Tech.* **43**(3), 19-26

Morgenroth E., Sherden T., van Loosdrecht M.C.M., Heijnen J.J., Wilderer P.A. (1997). Aerobic granular sludge in a sequencing batch reactor. *Wat. Res.* **31**(12), 3191-3194

Moy, B.Y.-P., Tay, J.-H., Toh, S.-K., Liu, Y., Tay, S.T.-L. (2002). High organic loading influences the physical characteristics of aerobic sludge granules. *Lett. Appl. Microbiol.* **34**(6), 407-412.

Peng D.C., Bernet N., Delgenes J.P., Moletta R. (1999). Aerobic granular sludge - A case report. *Wat. Res.* **33**(3), 890-893

Roessink, R., Eikelboom, D.H. (1997). Characterization of suspended solids in/out airlift biofilm reactors. *Wat. Sci. Technol.* **36**(1), 237-245.

Rosenwinkel K.H. (2000). *ATV-Handbuch Industrieabwasser / Lebensmittelindustrie (ATV-handbook industrial effluents / food industry).* 4th edition. Ernst und Sohn, Berlin

Shin H.S., Lim K.H., Park H.S. (1992). Effect of Shear-Stress on Granulation in Oxygen Aerobic Upflow Sludge Bed Reactors. *Wat. Sci. Tech.* **26**(3-4), 601-605

Tay J.H., Liu Q.S., Liu Y. (2001). The effects of shear force on the formation, structure and metabolism of aerobic granules. *Appl. Microbiol. Biotechnol.* **57**(1-2), 227-233

Tsuneda S., Ejiri Y., Nagano T., Hirata A. Formation mechanism of nitrifying granules observed in an aerobic upflow fluidized bed (AUFB) reactor. *Wat. Sci. Tech.* **49**(11-12), 27-34

Wilderer PA, Irvine RL, Goronszy MC (2001) Sequencing batch reactor technology. Scientific and technical report No.10. IWA-publishing, London

Yang S.-F., Liu Q.-S., Tay J.-H., Liu Y. (2004). Growth kinetics of aerobic granules developed in sequencing batch reactors. Lett. Appl. Microbiol. **38**(2), 106-112

Zhu, J., Wilderer, P.A. (2003). Effect of extended idle conditions on structure and activity of granular activated sludge. *Wat. Res.* **37**(9), 2013-2018

Modelling nutrient removal of an aerobic granular sludge lab-scale SBR using ASM3

M Lübken*, N Schwarzenbeck*, M Wichern*, PA Wilderer*

*Institute of Water Quality Control and Waste Management, Technical University of Munich, Am Coulombwall, 85748 Garching, Germany (E-mail: M.Luebken@bv.tum.de)

Abstract In a lab-scale SBR-reactor treating wastewater from the malting industry, aerobic granular sludge was cultivated. In this paper simulation results on the lab-scale SBR-reactor are given for the Activated Sludge Model no. 3 (Gujer et al 1999). ASM3 is a structured model for simulation of oxygen consumption, sludge production, nitrification and denitrification and was presented by the IWA task Group on Mathematical Modelling for Design and Operation of Biological Wastewater Treatment in 1999. It has been investigated if ASM3 could be used as a first simplification to simulate nutrient removal with aerobic granular sludge.

Keywords Activated Sludge Model No. 3, Modelling, SBR, Aerobic granular sludge

The Activated Sludge Model No. 3

The Activated sludge model No. 3 was published in 1999 by the IWA task Group on Mathematical Modelling for Design and Operation of Biological Wastewater Treatment (Gujer et al 1999). Improved possibilities to identify biological processes today have resulted in the development of the new model ASM3 to simulate nitrification, denitrification and degradation of COD. Compared with the ASM 1 (Henze et al 1987) the new model describes storage of organic substrates, the decay of heterotrophic organisms is modelled by the endogenous respiration and smaller anoxic yields for the heterotrophic organisms were used. The process of hydrolysis of slowly degradable COD (X_S) is not depending on redox conditions and is of less significance because the lysis process (with the production of slowly degradable COD) has been replaced by endogenous respiration. Hydrolysis of nitrogen has been combined with hydrolysis of COD. The decay rates for endogenous respiration of heterotrophic and autotrophic organisms are reduced under anoxic conditions. Values of kinetic and stoichiometric parameters of the calibrated version of ASM 3 by Koch et al 2000 have been used as initial parameters for the model. The main difference to the original publication (Gujer et al 1999) is that readily degradable COD is not measured by filtration but by respiration. Furthermore slowly degradable COD (X_S) includes soluble and particular components.

Lab-scale SBR-system set-up and operation

The modeled SBR-system is a lab reactor which was used by the Technical University of Munich to cultivate aerobic granular sludge. The daily influent amounted to 24 L. The volume of the reactor was 12 L with an internal diameter of 20 cm and a filling height of 38 cm. The volumetric exchange ratio was set to 2/3 that resulted in a hydraulic retention time (HRT) of 0.5 days. Aeration was provided by air bubble diffusors at a volumetric flow rate of about 800 L/h. Hereby, an average dissolved oxygen concentration of 8 mg/l could be achieved. Total cycle duration was 8h, with 6 min of fill, 120 min of anaerobic no mix conditions, 345 to 348 min of aeration, 5 to 2 min of settling and 4 min of effluent withdrawal. The temperature in the reactor during the simulation period was about 20 °C. The average sludge retention time was 4.5 days. Excess sludge wasn't removed separately, but corresponds with the amount of TSS in the effluent. A schematic depiction of the lab-scale reactor is given in Figure 1.

experimental set-up

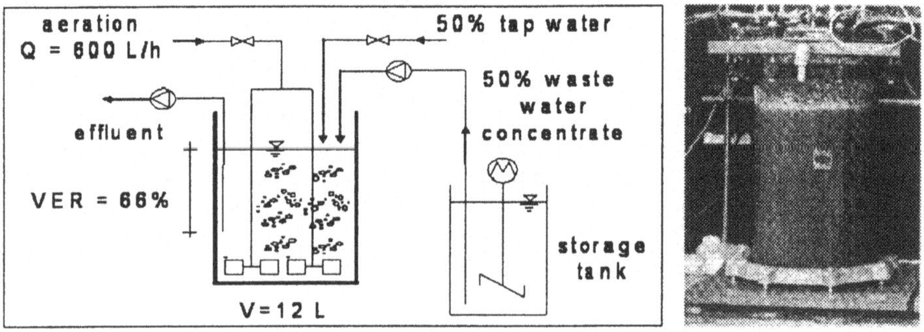

SBR-cycle:

total cycle	fill	anaerobic no-mix	aerobic	settle	decant
8 h	6 min	120 min	345-348 min	5-2 min	4 min

2h static fill (fill + anaerobic no-mix); classification of aggregates (settle + decant)

Figure 1: Schematic depiction of the modeled lab-scale SBR-reactor

In this reactor wastewater which is equivalent to that of the malting industry is treated. Substrate consists mainly of organic and particulate matter. The concentrations of COD, BOD_5 and TKN were adjusted according to characteristic values for malt house wastewaters, as they are given in the literature (Arndt 1999). The formation of aerobic granules was finished at the start of the simulation period. Granular sludge demonstrates extremely high settling velocities of microbial aggregates (Moy et al 2002) and a comparably higher and more robust metabolism (Zhu et al 2003). This leads to high removal efficiency in spite of high levels of organic loading rates as they are given in the lab-scale reactor. The average influent concentrations during the simulation period of COD_{total} and $COD_{dissolved}$ amounted to 1700 and 470 mg/l. The particulate fraction of COD_{total} in the feed had been raised to a maximum of 89%, which consisted of barley dust, plant cells and cellulose fragments (Schwarzenbeck et al 2004b). Furthermore, the average concentrations of TKN and NH_4-N in the influent during the simulation period were 45 and 3 mg/l, respectively.

Developed simulation model in SIMBA

The used simulation tool SIMBA in its version 4.1 supports simulation runs in a batch mode. In Figure 2 the developed SIMBA model for this paper is shown.

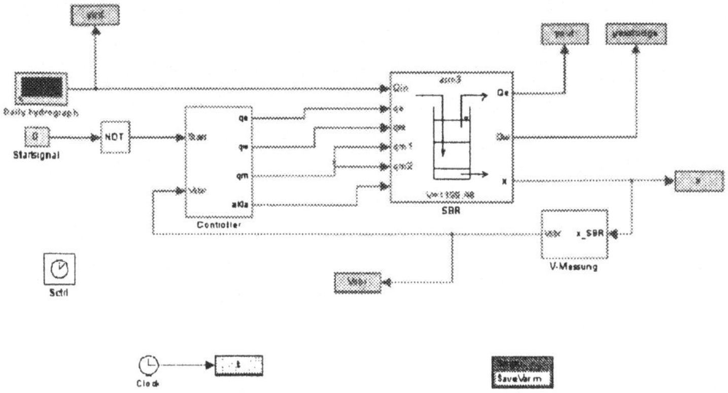

Figure 2: Depiction of the lab-scale SBR-reactor as a SIMBA simulation model

By the block SBR (sequencing batch reactor) the SBR process can be modelled, and by the subsystem controller the inputs of the SBR block are calculated and the start signal is given. To control a SBR process the cycle has to be defined. Within this subsystem the points in time are calculated at which a phase change takes place (Figure 3).

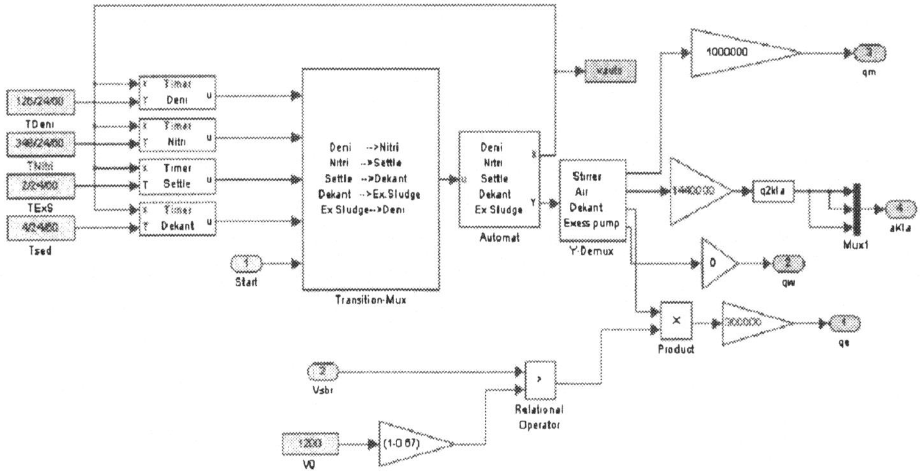

Figure 3: Depiction of the subsystem controller

Simulation results

Granular sludge occurred from 4/11/2002 to 18/12/2002. Focus of calibration was to model sludge production, oxygen demand and effluent values of the SBR. Dynamic concentrations in the SBR were not measured continuously, but, however, the amount of data made a dynamic simulation suitable. The lab-scale SBR-system is neither operated with enhanced biological phosphorous removal nor chemical precipitation. Therefore, the simulation could be started with ASM 3. Unfortunately, a detailed wastewater characterization as it is proposed in the Dutch guidelines for wastewater characterization in The Netherlands (ROELEVELD & VAN LOOSDRECHT 2002) hasn't been measured. However, wastewater COD fractions have been determined in the following way: S_I is based on the inert soluble COD in the effluent of the SBR. S_S is determined by subtracting the fraction S_I from the soluble COD in the influent (measured by 0.45 µm filter and assuming that X_S contains only particulate organic matter). The fraction of the heterotrophic biomass X_H was set to zero like it is proposed in the Dutch guidelines for wastewater characterization. The distribution of the remaining influent COD to the fraction of X_S and X_I was done iteratively by several simulation runs with a varying X_I and X_S content to predict the MLSS concentration within the SBR. At this, a higher X_I fraction results in a higher MLSS concentration. An overview of the average influent fractioning of individual model components is presented in Table 2.

Table 2: Fractionation of the organic matter

Definition	Symbol	Unit	Equation	% of COD
Inert soluble organic compounds	S_I	[$gCODm^{-3}$]	$COD_{dissolved,eff}$	4
Readily biodegradable organic compounds	S_S	[$gCODm^{-3}$]	$COD_{total,inf}-S_I$	7
Inert particulate organic compounds	X_I	[$gCODm^{-3}$]	iterative	26
Slowly biodegradable organic compounds	X_S	[$gCODm^{-3}$]	$COD_{total,inf}-X_I-X_H-S_S-S_I$	63
Biomass of heterotrophic organisms	X_H	[$gCODm^{-3}$]	$0*COD_{total,inf}$	0

The first simulation step was to predict the dissolved oxygen concentration in the model with the real situation and to calibrate the sludge retention time and the activated sludge concentration. Simulation results showed a high conformity with measured data. COD, ammonium and nitrate were modelled after the sludge age and the activated sludge concentration were calibrated. The simulation results of COD in the effluent are given in Figure 4.

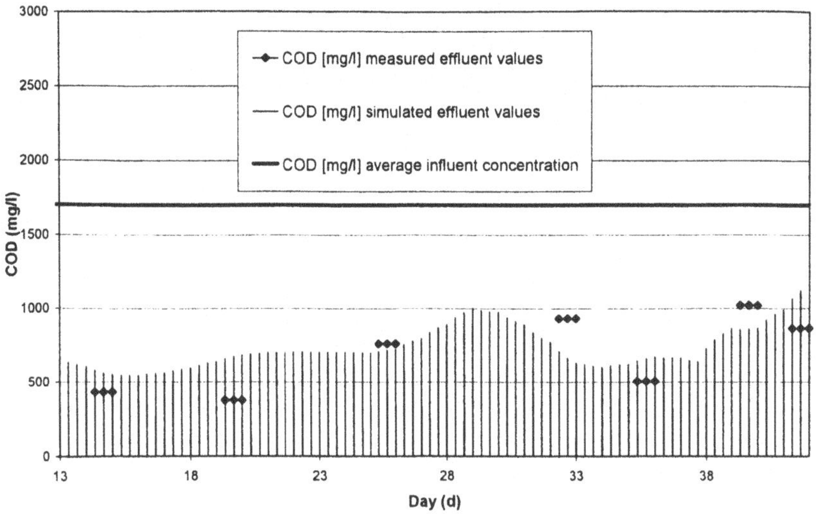

Figure 4: Results of the simulation and measured data of COD in the effluent

Figure 4 demonstrates high concordance between measured and simulated COD concentrations. The average COD can be sufficiently predicted by the model, concerning the dynamic course however, the amount of measurements is too small to make a statement possible. Normally, granular sludge has a higher COD removal efficiency, but as investigated by Schwarzenbeck et al 2004a the COD removal in this granular sludge system was comparable to suspended sludge systems. ASM3, which was developed to simulate systems using suspended sludge for wastewater treatment was able to predict the average COD removal of the existing granular sludge system.

In order to get a more detailed description of COD removal with granular sludge, a biofilm model is needed. Furthermore, the presence of a dense population of sessile protozoa, which could be identified responsible for uptake of particulate matter from the wastewater using tracer particles (Schwarzenbeck et al 2004b) is important for the reactor performance. However, as the COD removal was implicitly predicted by varying the X_I and X_S content of the influent COD and as the amount of data is quite small, no changes in the parameters of ASM3 were done to contribute to this topic. Further investigations for implementing the role of protozoa in the existing models are necessary.

In Figures 5 and 6 the simulation results for nitrate and ammonium are depicted.

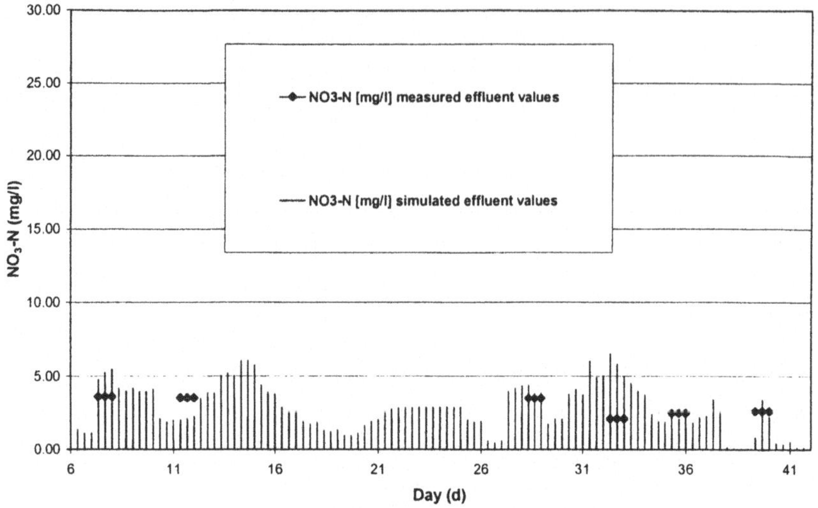

Figure 5: Results of the simulation and measured data of Nitrate in the effluent

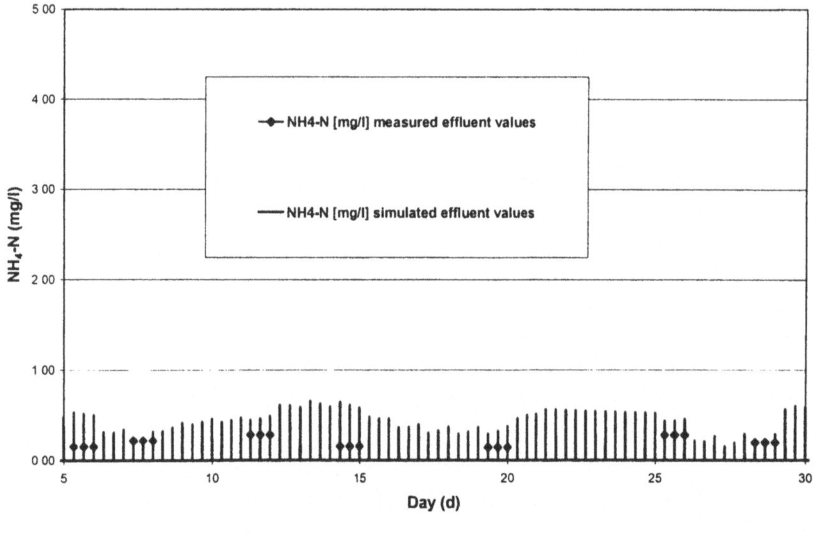

Figure 6: Results of the simulated and measured data of Ammonium in the effluent

Figures 5 and 6 present an agreement between measured and simulated values. Wastewater from the malting industry results mainly from the soaking process where a BOD concentration of 600 to 800 mg/l can be found (Arndt 1999). The BOD/TKN ratio between 13 and 18 for the present SBR-reactor is quite high. By the model a balance of nitrogen was carried out. The calculations showed that all the nitrogen of the influent is assimilated into the biomass. Therefore, denitrification and nitrification don't occur within the present reactor.

The behaviour of the existing SBR-system concerning COD, nitrate and ammonium removal could be modelled with the kinetic and stoichiometric parameters published by Koch et al 2000.

Conclusions

In this paper simulations with the Activated Sludge Model no. 3 (Gujer et al 1999) on a lab-scale SBR-reactor have been carried out for aerobic granular sludge. Information on the ASM3 model, the lab-scale SBR-reactor and its loading situation and the actual fractioning of COD used for simulation are presented. Furthermore, the development of a simulation model for SBR systems in SIMBA is shown. Kinetic and stoichiometric parameters of the calibrated version of ASM 3 by Koch et al 2000 have been used as parameters for the model. Despite the restriction that only flocculent sludge is considered in ASM3, the model proved capable of describing the performance of a lab-scale SBR-reactor concerning COD, ammonium and nitrate. However, for an in-depth analysis referring to the differences of suspended and granular sludge and especially for describing the characteristics of a single granule, biofilm processes should be included in the model. Concerning the role of protozoa and its integration in the existing models, further research is needed. For a detailed examination of the denitrification process it is evident to observe the role of storage products, like it was done by Beun et al 2001. All things considered, however, ASM3 can serve as a basis for modelling nutrient removal in an aerobic granular sludge system.

References

Alex J., et al. (2002). SIMBA 4.0 – Simulation of biological wastewater treatment. User's Guide.

Arndt G., (1999). Abwassersituation im Getränkebetrieb, in der Brauerei und Mälzerei (in German). *Brauwelt.* **30/31**, 1378-1379

Beun J.J., Heijnen J.J., van Loosdrecht M.C.M., (2001). N-removal in a granular sludge sequencing batch airlift reactor. *Biotech. & Bioeng.* **75**(1), 82-89

Beun J.J., van Loosdrecht M.C.M., Heijnen J.J. (2002). Aerobic granulation in a sequencing batch airlift reactor. *Wat. Res.* **36**(3), 702-712

De Kreuk M.K., de Bruin L.M.M. (2004). Aerobic Granule Reactor Technology. *Water and Wastewater Practitioner Series: STOWA Report*

Etterer T., Wilderer P.A. (2001). Generation and properties of aerobic granular sludge. *Wat. Sci. Tech.* **43**(3), 19-26

Gujer W., Henze M., Mino T., van Loosdrecht M.C.M., (1999). Activated Sludge Model No. 3. *Wat. Sci. Tech.* **39**(1), 183-193

Henze M., Grady C.P.L., Gujer W., Matsuo T., Marais G.V.R., (1987). Activated Sludge Model No. 1. *Scientific and Technical Report No. 1.* IAWPRC (now IWA)

Koch G., Kühni M., Gujer W., Siegrist H., (2000). Calibration and validation of Activated Sludge Model No. 3 for Swiss municipal wastewater. *Wat. Res.* **34**(14), 3580-3590

Lübken M., Wichern M., Rosenwinkel K.H., Wilderer P.A., (2003). Efficiency of different mathematical models for simulating enhanced biological phosphorus removal in activated sludge systems. *Invironmental Informatics Archives - ISEIS.*

Morgenroth E., Sherden T., van Loosdrecht M.C.M., Heijnen J.J., Wilderer P.A. (1997). Aerobic granular sludge in a sequencing batch reactor. *Wat. Res.* **31**(12), 3191-3194

Moy, B.Y.-P., Tay, J.-H., Toh, S.-K., Liu, Y., Tay, S.T.-L. (2002). High organic loading influences the physical characteristics of aerobic sludge granules. *Lett. Appl. Microbiol.* **34**(6), 407-412.

Peng D.C., Bernet N., Delgenes J.P., Moletta R. (1999). Aerobic granular sludge - A case report. *Wat. Res.* **33**(3), 890-893

Roeleveld P.J., van Loosdrecht M.C.M., (2002). Experience with guidelines for wastewater characterization in The Netherlands. *Wat. Sci. Tech.* **45**(6), 77-87

Roessink, R., Eikelboom, D.H. (1997). Characterization of suspended solids in/out airlift biofilm reactors. *Wat. Sci. Technol.* **36**(1), 237-245.

Schwarzenbeck N., Erley R., Mc. Swain B., Wilderer P.A., Irvine R.L., (2004a). Treatment of Malting Wastewater in a Granular Sludge Sequencing Batch Reactor (SBR). *Acta hydrochim. Hydrobiol.* **32**(1), 16-24

Schwarzenbeck N., Erley R., Wilderer P.A., (2004b). Aerobic granular sludge in an SBR-system treating wastewater rich in particular matter. *Wat. Sci. Tech.* **49**(11-12), 41-46

Shin H.S., Lim K.H., Park H.S. (1992). Effect of Shear-Stress on Granulation in Oxygen Aerobic Upflow Sludge Bed Reactors. *Wat. Sci. Tech.* **26**(3-4), 601-605

Tay J.H., Liu Q.S., Liu Y. (2001). The effects of shear force on the formation, structure and metabolism of aerobic granules. *Appl. Microbiol. Biotechnol.* **57**(1-2), 227-233

Wichern M., Lübken M., Blömer R., Rosenwinkel K.H. (2002). Efficiency of the Activated Sludge Model no. 3 for German wastewater on six different WWTPs. *Wat. Sci. Tech.* **47**(11), 211-218

Yang S.-F., Liu Q.-S., Tay J.-H., Liu Y. (2004). Growth kinetics of aerobic granules developed in sequencing batch reactors. Lett. Appl. Microbiol. **38**(2), 106-112

Zhu, J., Wilderer, P.A. (2003). Effect of extended idle conditions on structure and activity of granular activated sludge. *Wat. Res.* **37**(9), 2013-2018

Aerobic Granular Sludge - From Idea to Pilot Plant

M.K. de Kreuk, L.M.M. de Bruin and M.C.M. van Loosdrecht

Department of Biotechnology, Delft University of Technology, Julianalaan 67, 2628BC, Delft, The Netherlands (E-mail: m.dekreuk@tnw.tudelft.nl; m.c.m.vanloosdrecht@tnw.tudelft.nl)

Abstract During the last ten years, the aerobic granule technology has been developed from an idea and observation to a technology on its own that is ready to be scaled up and to be tested at pilot scale. Different mechanisms important for aerobic granulation have been described in the literature, e.g. yields of the involved organisms, shear and selection by settling rate. Analysis of a cross section of literature published in the last decade, shows that type of substrate, COD and N-load, superficial gas velocity or shear stress and oxygen concentration are important parameters. An important secondary parameter for the formation and maintenance of dense granules is the growth rate of the organisms that is influenced by cycle configuration or nitrogen load. Analyzing full-scale design aspects of a technology during laboratory research enhanced the development of this technology, which is currently evaluated at pilot scale. Bottlenecks as filling time, feast-famine regime and the use of an airlift could be eliminated in an early stage.

Keywords: Aerobic granular sludge, Scale-up, SBR, stability.

Introduction

Over the past ten years, the idea of a treatment plant solely consisting of aerobic granules developed into the first step towards reality: a pilot plant study. The benefits, expected from this technology are compact treatment plants and simple reactor design. The system will be based on Sequencing Batch Reactors (SBR's), allowing less sludge handling than in conventional activated sludge plants. This can be done, since the aerobic sludge technology is based on a one-reactor system; all processes, from influent feeding and aerobic and anaerobic conversion processes to separation of biomass and effluent withdrawal take place in one reactor. Because of the compact structure of the aerobic granules, high biomass concentrations can be obtained in these systems and therefore the volume load of these reactors can be high. The good settleability of the granules makes settling tanks superfluous (De Bruin et al., 2004; De Kreuk and De Bruin, 2004).

Because of the specific circumstances needed for the development of aerobic granular sludge, scaling-up this technology from an idea and laboratory experiments to a full-scale technology is not only a matter of economical reasoning but moreover a balance between economics, experimental findings and conventional scaling laws. This paper reviews the steps that are necessary in order to come to a pilot plant phase and focuses on the scale-up procedure as applied in the scope of the development of aerobic granular sludge technology, coupling practical solutions and economical advantages to conventional scale-up strategies.

Theory

Introduction in Granulation - Aspects

Granular sludge is well known in anaerobic systems and anaerobic granular sludge reactors (such as UASB, IC and EGSB) proved to perform effectively for over fifteen years. Much research has been carried out on factors concerning granule formation in anaerobic systems. One of the recurring explanations for this granulation is the occurrence of syntrophic juxtapositioned microcolonies. The hydrogen producing acetogens depend on the hydrogen consuming methanogens or sulphate reducing bacteria. Limiting the diffusion distances for hydrogen and other metabolic intermediate metabolites creates a favorable situation for the acetogens. Methanotrix is supposed to play an important role in the actual granule formation. This genus is present throughout all layers of anaerobic granules and has been observed in all types of anaerobic granules independent from the substrate composition. Therefore, it is often suggested that this organism plays an important role in granulation (Guiot et al., 1992; Fang et al., 1994).

In aerobic (but also in denitrifying) biofilms or granules syntrophic mechanisms are not at all or to a much lesser extent involved. This implies that other factors are important for granulation as well (Kosaric and Blaszcyk, 1990).

One of these other process factors is the selection of particles by their settling velocities. In an UASB reactor or other fluid-bed systems, particles are selected by their settling velocity, because only particles with settling velocities higher than the upward velocity of the fluid will stay in the reactor (Lettinga et al., 1980; Kosaric and Blaszcyk, 1990; Alphenaar et al., 1993).

The growth rate of the organisms, however, seems to be one of the main factors responsible for the density of granules or biofilms. Fast growing organisms will produce less dense granules than slow growing organisms (Villaseñor et al., 2000). For example, nitrifiers form a much denser biofilm than heterotrophs under the same circumstances. In anaerobic systems is reported that slow growing methanogens form denser biofilms than fast growing acidifying bacteria. Also, it is reported that an increase of the biomass surface-loading rate (i.e. the growth rate) decreases the biofilm density as well (Van Loosdrecht et al., 1995).

In aerobic biofilm reactors it has been found that shear force is an important factor for the formation of dense aggregates as well (Kwok et al., 1998). The biofilm density as well as the time to develop a fully covered carrier material decreased with a decreasing shear stress (Van Benthum et al., 1996a). This shows that shear effects for obtaining dense biofilms have to be balanced with detachment of biofilm fragments from the carrier. The main shear effect in a biofilm reactor results from particle/particle interaction, especially the collisions of bare carrier with the biofilms (Gjaltema et al., 1997, Kwok et al., 1998). Shear stress has also been recognized as an important factor in anaerobic systems. It has been reported that gas production causes enough movement and shear to decrease the average diameter of the anaerobic granules (Arcand et al., 1994).

Based on a wide range of experimental data a general hypothesis for the structure of biofilms was developed (Van Loosdrecht et al., 1995). This hypothesis states that the formation of dense and smooth biofilms occurs when the detachment rates are high compared to the biomass production. This hypothesis was also verified by mathematical modeling of the structure formation in biofilms (Picioreanu et al., 1999). Biofilms and granular sludge can be considered to be the same from a microbiological point of view, although there are obvious differences from a technical standpoint. The hypothesis stated above might be helpful in explaining the conditions required for the formation of good granular sludge. This will be further elaborated on in the following part.

Aerobic granulation – an overview

From the parallels between the aerobic and anaerobic research on granular sludge, as briefly described above, it can be stated that there are three main factors, which determine the granulation process. These are the growth rates of the organisms involved, the shear applied to the granules and the selection via settling rate in the reactor. Especially the last two parameters are very important for the design of a granular sludge reactor. Differences between aerobic and anaerobic granules seem to depend more on these aspects than on the specific microbial populations involved. The aerobic granule research performed during the last decade focused on gaining knowledge about factors necessary for granule formation and conversion processes on laboratory scale.

The research performed in our group, leading to aerobic granular sludge technology started in the early nineties with studies on morphology of microbial growth in biofilms, flocs and particles (Tijhuis et al., 1994; Gjaltema et al., 1994; Tijhuis et al. 1995; Van Loosdrecht et al., 1995; Van Benthum et al., 1996a; Van Benthum et al., 1996b; Gjaltema et al., 1997; Picioreanu, 1999; Beun et al., 1999). In continuously operated systems, it was possible to grow granular sludge with slowly biodegradable substrates (e.g. methanol or ammonia). In the case of easily biodegradable substrates, these systems can only be used when enough basalt is added to the system (as carrier material and for shear reasons). These continuously operated reactors are amongst others known as CIRCOX technology (Frijters et al. 1997). Further research concerning the formation of storage polymers (Beun et al., 2002b) finally resulted in the idea of growing aerobic granules without carrier material on readily biodegradable substrates in a Sequencing Batch Reactor (Morgenroth et al., 1997; Beun et al., 1999; Dangcong et al. 1999). The conversion of readily biodegradable COD into a substrate yielding a lower maximal growth rate facilitated granule formation. In 1998, an international patent was submitted and granted (Heijnen and Van Loosdrecht, 1998). An extension of this first patent was submitted in 2004, including the description of anaerobic feeding (Van Loosdrecht and De Kreuk, 2004).

An increase of the research intensity on aerobic granulation was observed since the end of the nineties. However, already in 1991 a pilot study for an aerobic upflow sludge blanket reactor, operated as an UASB, was established, but with pure oxygen injection into the influent (Mishima and Nakamura, 1991). In this reactor operated with low shear and high oxygen concentrations, aerobic granules of 2-8 mm diameter were observed, with an SVI of 72 ml g^{-1} (table 1).

An overview of a broad selection of different research papers on aerobic granule formation is given in table 1. The main topics of interest have been the relations of granule size, morphology, SVI, density, hydrophobicity and conversion processes under different process conditions. The earliest papers focused on general aspects of granule formation (numbers 1-7 in table 1: Mishima and Nakamura, 1991; Tijhuis et al, 1994; Van Benthum et al, 1996a; Morgenroth et al., 1997; Dancong et al., 1999; Beun et al, 1999; Beun et al. 2000). The specific conditions that were further investigated were:

- types of substrate (numbers 8-11 and 18 in table 1: Tay et al., 2001b; Tay et al., 2002; Moy et al., 2002; Jiang et al., 2002; Jang et al., 2003; Schwarzenbeck et al., 2004);
- method substrate was dosed (numbers 12, 16, 17 in table 1: Dulekurgen et al., 2003; De Kreuk and Van Loosdrecht, 2004; McSwain et al., 2004);

- shear stress (by adjusting the superficial gas velocity) (number 19 in table 1: Tay et al., 2001a);
- dissolved oxygen concentration (by recycling the off-gas) (numbers 16 and 20 in table 1: De Kreuk and Van Loosdrecht, 2004; Mosquera-Corral et al., submitted);
- organic load (numbers 10 and 13 in table 1: Jiang et al., 2002; Liu et al., 2003a);
- COD/N ratio (numbers 10, 13 and 14 in table 1: Jiang et al., 2002; Liu et al., 2003a; Liu et al., 2003b);
- settling velocity (number 15 in table 1: Qin et al., 2004a);
- selection for slow growing organisms (numbers 12, 14 and 16 in table 1: Dulekurgen, 2003; Liu et al., 2004a; De Kreuk and Van Loosdrecht, 2004).

An overview of these and other conditions (e.g. Ca^{2+} concentration, hydraulic retention times, seed sludge and inhibition) believed to be important for aerobic granule formation is given by Liu and Tay (2004).

The results presented in the research summarized in table 1 were analyzed by determining correlations between granule diameter, SVI, suspended solid concentration or granule density, and certain process parameters. Some of the process parameters (such as superficial gas velocity, minimal settling velocity of the granules and COD/N ratio) had to be calculated from the description in the method section of the cited papers. By comparing all results in table 1, a clear relation between the SVI and density could be observed. A decreasing granule density leads to an increased SVI, which leads to a lower suspended solids concentration in the reactor. No trends were observed between granule diameter and SVI, density or suspended solids concentration. Other relations found by analyzing all parameters are summarized in table 2. Although the relations also depend on other parameters than the compared ones, table 2 gives an idea which parameters are important and which are less important in granule formation.

No trends were observed between the analyzed parameters and the suspended solids concentration. Reported suspended solids concentrations fluctuated from 0.88 to 16.2 g l^{-1} (average value was 6.8 g l^{-1}). The type of substrate influenced the analyzed parameters to a large extend. Glucose led to the largest granules, while acetate resulted in the lowest SVI and the highest density. The COD-load did not give a clear trend, but a small increase in granule diameter and decrease in density was observed at increased COD values. Moy et al. (2002) reports the disintegration of granules at acetate loads above 9 kg COD m^{-3} d^{-1}. In other studies 9 kg COD m^{-3} d^{-1} was the highest load tested, while most studies were carried out at 6 kg COD m^{-3} d^{-1} or less. The granule strength was reported to decrease at increasing load (Liu et al. 2003a). This is in line with the observation of disintegration at high loads. However, it is expected that with decreased granule strength, the diameter of the granules would be negatively effected by breaking of the granules. However, granule diameter increases with increasing load, so probably the low density cores of the granules still have enough strength to stay intact.

The influence of COD/N ratio (or increased N-load) demonstrated a similar trend on granule characteristics as the COD load. Increased COD/N results in an increased diameter and a decreased density combined with a slightly increased SVI. Since the substrate influences these parameters as well, the influence of COD/N ratio was also solely studied for the experiments with acetate, but this resulted in the same outcome. A decreased COD/N or increased N-load leads to favorable conditions for slow growing autotrophic organisms. Analogous to biofilm development, this will lead to denser granules. This phenomenon is observed with increased nitrogen load (Liu et al., 2004), with the selection for slow growing phosphate accumulating organisms (Dulekurgen et al., 2003; De Kreuk and Van Loosdrecht, 2004) and with the influence of feast-famine regime (McSwain et al., 2004; Mosquera-

Corral et al, Submitted). In relation with biofilm formation, growth rate seems to play a major role in the formation of aerobic granular sludge.

Surprisingly, no trends were observed between the studied parameters and the settling time or minimal settling velocity of the granules. Settling times between 0.7 and 33 m h^{-1} were used for the different studies and within this range, settling time did not influence granule characteristics when all studies were compared.

A more decisive parameter for granule diameter was the superficial gas velocity. The trends displayed a maximal granule diameter at superficial gas velocity 0.02 m s^{-1}. At lower and higher gas velocities the granule diameter was smaller, indicating a need for shear, but also demonstrating the destruction of granules at high shear stress. For SVI and solids concentration no clear trends were observed. Unfortunately, biofilm density results are only reported for experiments at 0.025 m s^{-1} superficial gas velocity, which ranged from 12 to 78 g l^{-1} under these circumstances.

The influence of the dissolved oxygen concentration is difficult to interpret. Most of the cited papers only report values larger than 2 or 3 mg l^{-1} and do not give exact values. However, the experiments at low oxygen saturation (5, 19 and 20, oxygen concentration respectively <1.0 mg l^{-1}, 1.5 mg l^{-1} and app. 4 mg l^{-1}) describe the agglomeration of granules to flocs after stopping the aeration (Mishima and Nakamura, 1991; Mosquera-Corral, submitted) as well as formation of new flocs (Tay et al, 2001a; Tay et al., 2004). Tay et al., (2001a) related the absence of the granules to lack of shear stress and Liu and Tay (2004) report that the DO concentration is not a decisive parameter in granule formation. This highly contradicts the experimental findings when low oxygen concentrations were used. The influence of dissolved oxygen concentration should not be underestimated in the design of full-scale treatment plants.

In summary, it can be said that the main investigated aspects, which have been investigated in relation to granule formation, are type of substrate in combination with the organic and the nitrogen load, the shear effects and the oxygen concentration. A secondary positive effect of the process parameters comes from selection of slow growing organisms by changing cycle configuration and/or COD/N ratio.

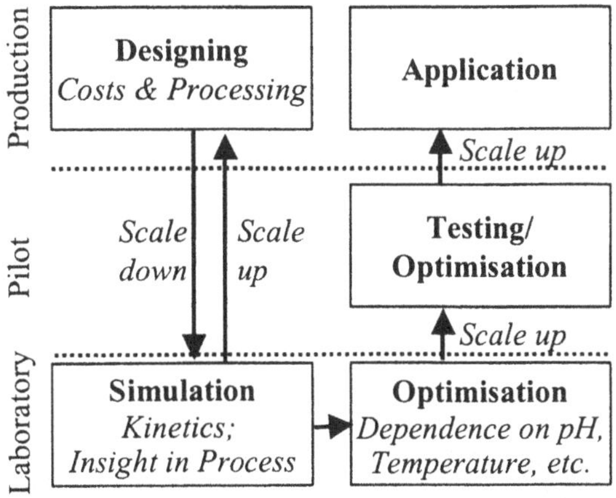

Figure 1: Approach for the design of a new system based on aerobic sludge technology

Design of an Aerobic Granular Sludge System – Scale Up

Approach

As described above, aerobic granulation technology for the application in wastewater treatment has been demonstrated, so far, only at laboratory scale. The full-scale experience may be derived from comparable processes such as large scale SBRs for wastewater treatment (Wilderer et al, 2000), continuously fed biofilm airlift reactors (Heijnen et al., 1993; Frijters et al., 1997) and EBPR systems (Van Loosdrecht et al., 1997). Because of the unique process conditions for selecting and growing aerobic granules, this process equipment cannot be applied one-to-one to aerobic granular sludge systems.

To overcome the existing knowledge deficit and to design a system that is able to compete with existing technology, a combined step-by-step and scale-up/scale-down approach has been used for the final large-scale design. This is schematically shown in Figure 1; experimental results were translated into design aspects for a full-scale plant and bottlenecks from the design were investigated with laboratory scale experiments (scale-up/scale-down). Consequently, the design of a full-scale installation occurred at the same time as the laboratory experiments. With knowledge of most bottlenecks and essential processes, the process could be optimized after which pilot-scale experiments could be carried out. When this phase succeeds, the step to a full-scale application could be made (step-by-step scaling).

Bottlenecks

From the large-scale design, several bottlenecks were found for large-scale application. The investigated factors influencing reactor design were: pulse feeding, using an airlift reactor in combination with the expected influent flows and column height, rainwater versus dry weather flow, post- or pre-treatment and the decanting time in relation to the post-treatment step. The bottlenecks related to granule stability were amongst others the oxygen concentration during aeration, feeding pattern, temperature, adequate nutrient removal (COD, N and P) and type of substrate. Most bottlenecks are or will be described in other research papers (De Bruin et al, 2004; De Kreuk and Van Loosdrecht, 2004; Mosquera-Corral et al., submitted; De Kreuk et al., In Preparation). As an illustration two bottlenecks will be further highlighted below.

Pulse feeding the aerobic granular sludge reactor, as applied in the first laboratory scale experiments, turned out to be economically and technologically very unattractive at large-scale. This pulse feed was needed for the creation of a feast-famine regime; high substrate concentrations at the beginning of each cycle are needed for maximum conversion of readily biodegradable substrate into storage polymers (Beun et al., 2002a; De Kreuk and Van Loosdrecht, 2004). However, a pulse feed would require buffer tanks for storage of the continuous influent flow received by a wastewater treatment plant. Since the advantages of aerobic granule technology are mainly based on compactness of the installation, the use of buffer tanks is highly unfavorable. Furthermore, pulse-feeding would require a large pumping capacity and a large aeration capacity to supply oxygen during feeding. To overcome this problem, substrate has to be dosed to the system in such way that storage of the C-source is accomplished and a continuously incoming flow can be fed alternating to the different SBR's operated in parallel. This means that feeding N reactors in $1/N^{th}$ of the cycle time can be translated in a continuous influent flow to the total installation.

The demand that arose from the large-scale design was to increase the feeding time from the original experimental pulse feed (3 minutes) to at least one hour. The solution for this problem found at laboratory scale was to feed the reactor anaerobically. The influent was dosed from the bottom of the reactor passing the settled bed of granules. This alternating anaerobic feeding period and aerobic reaction period led to conditions favoring the growth of phosphate accumulating organisms. This selection method turned out to be very advantageous for granule stability as well as nutrient removal (De Kreuk and Van Loosdrecht, 2004). As a conclusion it can be said that by solving a technological large-scale bottleneck by means of a small-scale experiment (scale-up/scale-down), the overall aerobic granular sludge process has been highly improved.

Another parameter sensitive to the costs and the design of a post treatment step is the decanting time (De Bruin et al., 2004). Short decanting times are favorable for the total cycle duration and thus for the reactor size, whereas large decanting times reduces the flow to the post-treatment, leading to a smaller post-treatment installation. The solution for this bottleneck could be a simultaneous influent feeding and effluent extraction phase, which would shorten the decanting period and the total cycle duration and would cause a continuous minimal flow to the post-treatment plant. Simultaneous feeding-decanting could be achieved by feeding the influent through the settled bed, without fluidizing the particles. If granules are in a fixed state (settled bed), plug-flow conditions are prevailing and 90% of the effluent present can be exchanged simultaneously with the incoming influent (Arnz et al., 2000). Theoretical modeling demonstrated that an applied influent superficial velocity of 4 mg h^{-1} would not fluidize the granules if the bed voidage is higher than 0.6 and the particle sphericity is higher than 0.7. However, tests in laboratory reactors (operated as in de Kreuk and Van Loosdrecht, 2004) showed fluidization at 0.75 mg h^{-1} with this void fraction and sphericity. The minimum fluidization flow will set the geometrical design for the full-scale reactor, because the maximum influent flow and the minimum fluidization velocity determine the diameter of the reactor. However, wall effects at laboratory scale are relatively large, so the experimental values found are expected not to be representative for full-scale applications. Furthermore, the granule morphology in the laboratory scale experiments (acetate as single substrate) might be different from granules grown at presettled sewage. So, before a decision can be made about simultaneous feeding-decanting, minimum fluidization has to be further investigated at pilot-scale, where wall effects are lower (step-by-step scaling method).

Conclusions

Analysis of a cross section of literature published in the last ten years of aerobic granular sludge research reveals the process parameters that are important for granule formation: type of substrate, COD and N-load, shear stress and concentration profiles. These parameters can be translated to the growth rate as (most) important secondary parameter for formation of dense and stable granules.

A scale-up/scale-down approach in the early stage of the development of a new technology is considered very useful to direct the laboratory research towards solving full-scale bottlenecks. When full-scale design and laboratory-scale processes are optimized, a step-by-step approach (pilot-plant) is needed to come to the final full-scale technology. Thus, constant awareness about the consequences of results obtained at laboratory- and pilot-scale for the final application enhances the development of a new technology and increases the chance of a large-scale success.

Acknowledgements

This research was funded by the Dutch Foundation for Water Research (STOWA, TNW99.262) and STW (DPC5577) within the framework of the "aerobic granule reactors" project. The project is in close co-operation with DHV-Water in Leusden, The Netherlands

References

Alphenaar, P.A., Visser A. and Lettinga G. (1993). The Effect of Liquid Upward Velocity and Hydraulic Retention Time on Granulation in Uasb Reactors Treating Wastewater with a High Sulphate Content. Bioresource Technology. 43, 249-258.

Arcand, Y., Guiot S.R., Desrochers M. and Chavarie C. (1994). Impact of the Reactor Hydrodynamics and Organic Loading on the Size and the Activity of Anaerobic Granules. The chemical engineering journal. 56, B23 - B35.

Arnz, P., Esterl M.S., Nerger C., Delgado A. and Wilderer P.A. (2000). Simultaneous Loading and Draining as a Means to Enhance Efficacy of Sequencing Biofilm Batch Reactors. Wat. Res. 34(5), 1763 - 1766.

Beun, J.J., Hendriks A., van Loosdrecht M.C.M., Morgenroth M., Wilderer P.A. and Heijnen J.J. (1999). Aerobic Granulation in a Sequencing Batch Reactor. Wat. Res. 33(10), 2283-2290.

Beun, J.J., van Loosdrecht M.C.M. and Heijnen J.J. (2000). Aerobic Granulation. Wat. Sci. Technol. 41(4-5), 41-48.

Beun, J.J., van Loosdrecht M.C.M. and Heijnen J.J. (2002a). Aerobic Granulation in a Sequencing Batch Airlift Reactor. Wat. Res. 36(3), 702-712.

Beun, J.J., Dircks K., Van Loosdrecht M.C.M. and Heijnen J.J. (2002b). Poly-B-Hydroxybutyrate Metabolism in Dynamically Fed Mixed Microbial Cultures. Wat. Res. **36**(5), 1167-1180.

Dangcong, P., Bernet N., Delgenes J.-P. and Moletta R. (1999). Aerobic Granular Sludge--a Case Report. Wat. Res. 33(3), 890-893.

De Bruin, L.M.M., De Kreuk M.K., van der Roest H.F.R., Van Loosdrecht M.C.M. and Uijterlinde C. (2004). Aerobic Granular Sludge Technology, Alternative for Activated Sludge Technology? Wat. Sci. Technol. 49(11-12), 1-9.

De Kreuk, M.K. and De Bruin L.M.M. (2004). Aerobic Granule Reactor Technology. London, IWA Publishing.

De Kreuk, M.K., Pronk M. and Van Loosdrecht M.C.M. (In Preparation). Temperature Effects on Aerobic Granulation Process.

De Kreuk, M.K. and Van Loosdrecht M.C.M. (2004). Selection of Slow Growing Organisms as a Means for Improving Aerobic Granular Sludge Stability. Wat. Sci. Technol. 49(11-12), 9-19.

Dulekgurgen, E., Ovez S., Artan N. and Orhon D. (2003). Enhanced Biological Phosphate Removal by Granular Sludge in a Sequencing Batch Reactor. Biotechnology Letters. 25(9), 687 - 693.

Fang, H.H.P., Chui H.K. and Li Y.Y. (1994). Microbial Structure and Activity of Uasb Granules Treating Different Wastewaters. Wat. Sci. Technol. 30(12), 87 - 96.

Frijters, C.T.M.J., Eikelboom D.H., Mulder A. and Mulder R. (1997). Treatment of Municipal Wastewater in a Circox® Airlift Reactor with Integrated Denitrification. Wat. Sci. Technol. 36(1), 173-181.

Gjaltema A, Arts PAM, Van Loosdrecht MCM, Kuenen JG, Heijnen JJ. (1994). Heterogeneity of Biofilms in Rotating Annular Reactors: Occurrence, Structure, and Consequences. Biotechnol. Bioeng. 44(2):194-204.

Gjaltema, A., Vinke J.L., Van Loosdrecht M.C.M. and Heijnen J.J. (1997). Abrasion of Suspended Biofilm Pellets in Airlift Reactors: Importance of Shape, Structure and Particle Concentrations. Biotechnol. Bioeng. 53, 88 - 99.

Guiot, S.R., Pauss A. and Costerton J.W. (1992). A Structured Model of the Anaerobic Granule Consortium. Wat. Sci. Technol. 25(7), 1 - 10.

Heijnen, J.J., Loosdrecht M.C.M., Mulder R., Weltevrede R. and Mulder A. (1993). Development and Scale-up of an Aerobic Biofilm Air-Lift Suspension Reactor. Wat. Sci. Technol. 27(5-6), 253-261.

Heijnen, J.J. and Van Loosdrecht M.C.M. (1998). Method for Acquiring Grain-Shaped Growth of a Microorganism in a Reactor. Patent Cooperation Treaty (PCT). US, European patent, Technische Universiteit Delft: 13.

Jang, A., Yoon Y.-H., Kim I.S., Kim K.-S. and Bishop P.L. (2003). Characterization and Evaluation of Aerobic Granules in Sequencing Batch Reactor. J. Biotechnol. 105(1-2), 71-82.

Jiang, H.L., Tay J.H. and Tay S.T.L. (2002). Aggregation of Immobilized Activated Sludge Cells into Aerobically Grown Microbial Granules for the Aerobic Biodegradation of Phenol. Letters in Applied Microbiology. 35(5), 439-445.

Kosaric, N. and Blaszcyk R. (1990). Microbial Aggregates in Anaerobic Waste Water Treatment. Adv. Biochem. Eng./Biotechnol. 42, 27 - 62.

Kwok, W.K., Picioreanu C., Ong S.L., van Loosdrecht M.C.M., Ng W.J. and Heijnen J.J. (1998). Influence of Biomass Production and Detachment Forces on Biofilm Structures in a Biofilm Airlift Suspension Reactor. Biotechnol. Bioeng. 58(4), 400- 407.

Lettinga, G., Van Velsen A.F.M., Hobma S.W., De Zeeuw W. and Klapwijk A. (1980). Use of the Upflow Sludge Blanket (Usb) Reactor Concept for Biological Wastewater Treatment, Especially for Anaerobic Treatment. Biotechnol. Bioeng. 22, 699 - 734.

Liu, Q.S., Tay J.H. and Liu Y. (2003a). Substrate Concentration-Independent Aerobic Granulation in Sequential Aerobic Sludge Blanket Reactor. Environmental Technology. 24(10), 1235-1242.

Liu, Y. and Tay J.-H. (2004). State of the Art of Biogranulation Technology for Wastewater Treatment. Biotechnology Advances. 22(7), 533-563.

Liu, Y., Yang S.F. and Tay J.-H. (2003b). Elemental Compositions and Characteristics of Aerobic Granules Cultivated at Different Substrate N/C Ratios. Appl. Microbiol. Biotechnol. 61(5-6), 556 - 561.

Liu, Y., Yang S.-F. and Tay J.-H. (2004). Improved Stability of Aerobic Granules by Selecting Slow-Growing Nitrifying Bacteria. J. Biotechnol. 108(2), 161-169.

McSwain, B.S., Irvine R.L. and Wilderer P.A. (2004). The Effect of Intermittent Feeding on Aerobic Granule Structure. Wat. Sci. Technol. 49(11), 19-25.

Mishima, K. and Nakamura M. (1991). Self-Immobilization of Aerobic Activated Sludge - a Pilot Study of the Aerobic Upflow Sludge Blanket Process in Municipal Sewage Treatment. Wat. Sci. Technol. 23, 981 - 990.

Morgenroth, E., Sherden T., van Loosdrecht M.C.M., Heijnen J.J. and Wilderer P.A. (1997). Aerobic Granular Sludge in a Sequencing Batch Reactor. Wat. Res. 31(12), 3191 - 3194.

Mosquera-Corral, A., De Kreuk M.K., Heijnen J.J. and Van Loosdrecht M.C.M. (Submitted). Effects of Dissolved Oxygen on N-Removal in an Aerobic Granular Sludge Reactor. Wat. Res.

Moy, B.Y.P., Tay J.H., Toh S.K., Liu Y. and Tay S.T.L. (2002). High Organic Loading Influences the Physical Characteristics of Aerobic Sludge Granules. Letters in Applied Microbiology. 34, 407-412.

Picioreanu, C., Van Loosdrecht M.C.M. and Heijnen J.J. (1999). Discrete-Differential Modelling of Biofilm Structure. Wat. Sci. Technol. 39(7), 115-122.

Qin, L., Liu Y. and Tay J.-H. (2004a). Effect of Settling Time on Aerobic Granulation in Sequencing Batch Reactor. Biochemical Engineering Journal. 21(1), 47-52.

Qin, L., Tay J.H. and Liu Y. (2004b). Selection Pressure Is a Driving Force of Aerobic Granulation in Sequencing Batch Reactors. Process Biochemistry. 39(5), 579-584.

Schwarzenbeck, N., Erley R. and Wilderer P.A. (2004). Aerobic Granular Sludge in an Sbr-System Treating Wastewater Rich in Particulate Matter. Wat. Sci. Technol. 49(11-12), 41-46.

Tay, J.H., Liu Q.S. and Liu Y. (2001a). The Effects of Shear Force on the Formation, Structure and Metabolism of Aerobic Granules. Appl. Microbiol. Biotechnol. 57, 227 - 233.

Tay, J.H., Liu Q.S. and Liu Y. (2001b). Microscopic Observation of Aerobic Granulation in Sequential Aerobic Sludge Blanket Reactor. Journal of Applied Microbiology. 91(1), 168-175.

Tay, J.H., Liu Q.S. and Liu Y. (2002). Characteristics of Aerobic Granules Grown on Glucose and Acetate in Sequential Aerobic Sludge Blanket Reactors. Environmental Technology. 23(8), 931-936.

Tay, J.H., Liu Q.S. and Liu Y. (2004). The Effect of Upflow Air Velocity on the Structure of Aerobic Granules Cultivated in a Sequencing Batch Reactor. Wat. Sci. Technol. 49(11-12), 35-40.

Tijhuis L, Rekswinkel E, van Loosdrecht MCM, Heijnen JJ. (1994). Dynamics of Population and Biofilm Structure in the Biofilm Airlift Suspension Reactor for Carbon and Nitrogen Removal. Wat. Sci. Technol. 29(10-11):377-384.

Tijhuis L, Hijman B, Van Loosdrecht MCM, Heijnen JJ. (1995). Influence of Detachment, Substrate Loading and Reactor Scale on the Formation of Biofilms in Airlift Reactors. Appl. Microbiol. Biotechnol. 45:7-17.

Van Benthum, W.A.J., Garrido-Fernandez J.M., Tijhuis L., van Loosdrecht M.C.M. and Heijnen J.J. (1996a). Formation and Detachment of Biofilms and Granules in a Nitrifying Biofilm Airlift Suspension Reactor. Biotechnel. Prog. 1996(12), 764-772.

Van Benthum WAJ, Van Loosdrecht MCM, Heijnen JJ. (1996b). Control of Heterotrophic Layer Formation on Nitrifying Biofilms in a Biofilm Airlift Suspension Reactor. Biotechnol. Bioeng. 53(4):397 - 405

Van Loosdrecht M.C.M. Eikelboom D. H., Gjaltema A., Mulder A., Tijhuis L., Heijnen J. J. (1995). Biofilm Structures. Wat. Sci. Technol. 32(8), 35-43.

Van Loosdrecht, M.C.M. and De Kreuk M.K. (2004). Method for the Treatment of Waste Water with Sludge Granules. Bureau voor Industriele Eigendom. The Netherlands, International, Technische Universiteit Delft.

Van Loosdrecht, M.C.M., Hooijmans C.M., Brdjanovic D. and Heijnen J.J. (1997). Biological Phosphate Removal Processes. Appl. Microbiol. Biotechnol. 48(3), 289-296.

Villaseñor, J.C., van Loosdrecht M.C.M., Picioreanu C. and Heijnen J.J. (2000). Influence of Different Substrates on the Formation of Biofilms in a Biofilm Airlift Suspension Reactor. Wat. Sci. Technol. 41(4-5), 323-330.

Wilderer, P.A., Irvine R.L. and Goronszy M., C. (2000). Sequencing Batch Reactor Technology. Colchester, UK, IWA Publishing.

Table 1: Overview of published data about aerobic granular sludge concerning growth conditions, diameter, SVI and suspended solid concentration.

Reference	Diameter (mm)	SVI (ml g^{-1})	SS (g l^{-1})	Remarks	Growth conditions
1. Mishima and Nakamura (1991)	2-8	41-143	8.2	-	Upflow reactor (continuous); 14°C; DO 63 mg l^{-1} (pure O2 injection); BOD-load 1.57 (sewage), BOD/N=3.7
2. Tijhuis et al. (1994)	0.35	n.a.	0.7	Biomass density: 20; granules washed out	Biofilm Airlift Reactor (continuous); Carrier material; 28.5°C; v_{gs} 2; COD-load 5.0 (acetate), COD/N=9
3. Van Benthum et al. (1996a)	> 0.3	n.a.	25-30	Nitrifying biomass density: 70;	Biofilm Airlift Reactor (continuous); Carrier material; 30°C; v_{gs} 2.3; bicarbonate as C-source; N-load 5
4. Morgenroth et al. (1997)	2.35	n.a.	0.88	Solid, but soft texture (fungi)	SBR; DO>2 mg l^{-1}; $v_{p,min}$ 30; COD load 0.72 (molasses), COD/N=18
5. Dangcong et al. (1999)	0.3 - 0.5	80-100	4 - 4.5	Agglomeration into flocs when aeration stopped	SBR; 25 °C; DO <1.0 mg l^{-1}; stirring at 400 rpm; COD-load 1.5 (acetate), COD/N=13
6. Beun et al. (1999)	3.3	124-156	3.2	Granule density: 12	SBR; 20 °C; DO 100%; v_{gs} 2.5; $v_{p,min}$ 12; COD load 5 (ethanol), COD/N=26
7. Beun et al. (2000)	1.0	70	4	Granule density: 48	SBR (airlift); 20 °C; DO 100% (v_{gs} 2.5); $v_{p,min}$ 16.2; COD-load 2.3 (acetate), COD/N=14
8. Tay et al (2001b) and Tay et al. (2002)	a: 1.1 g: 2.4	a: 50-80 g: 51-85	a: 8.6 g: 5.9	Granule density: 32.2^{a}, 41.1^{g}; filamentous^{g}	SBR, v_{gs} 2.5; $v_{p,min}$ 30-36; COD-load 6 (a=acetate, g=glucose), COD/N=38
9. Moy et al. (2002)	a: 1.96; 4.2 g: 2.7; 2.95; 3.06; 3.30	a: 49; 42 g: 106; 85; 74; 31	n.a.	a: disintegration COD>9 g: SVI decreased at increased load	SBR; 25 °C; v_{gs} 2; $v_{p,min}$ 1.5; COD-load resp. $6^{g,a}$, $9^{g,a}$ 12^{g} and 15^{g} (a=acetate, g=glucose), COD/N=$38^{g,a}$ $56^{g,a}$, 75^{g}, 94^{g}

10. Jiang et al. (2002)	0.35-0.6	40	7	After adaptation all phenol was degraded	SBR; 25 °C; v_{gs} 3; $v_{p,min}$ 1.2 – 7.2; COD-load 2.68 (phenol), COD/N=21
11. Jang et al. (2003)	1-1.3	70-90	6	AOB in outer layers (FISH)	SBR; 25 °C; DO>2 mg l^{-1} (v_{gs} 0.4) ; $v_{p,min}$ 0.7; COD-load 1.3 (glucose & acetate), COD/N=21
12 Dulekurgen et al. (2003)	2.5-3	45	4.4	EBPR performance	SBR; anaerobic feeding (2h), followed by aeration period (3h); COD-load 1.2, (acetate) COD/N=13.6
13. Liu et al. (2003a)	1.6; 1.8; 1.8; 1.9	41; 43; 36; 34	8.4; 9.5; 11.2, 12.3	lower granule strength at higher load	SBR; 25 °C; $v_{p,min}$ 13; v_{gs} 2.4; COD-load 1.5, 3, 6 and 9 (acetate), COD/N=5.5; 11; 22; 33
14. Liu et al. (2003b) and Liu et al. (2004)	0.4; 0.5; 1.5; 1.9	app. 59;62; 68; 77	n.a.	lower hydrophobicity at higher COD/N	SBR; 25 °C; DO > 2 mg l^{-1}, COD-load 3 (ethanol), COD/N= 3.3; 5; 10; 20
15. Qin et al. (2004a) and Qin et al. (2004b)	10; 15; 35; 100%>0.35	app. 145; 115; 75; 50	5.3; 4.9; 5.5; 5.4	lower SVI with higher hydrophobicity	SBR, v_{gs} 2.5; DO>50%; $v_{p,min}$ 1.8; 2.4; 3.6; 7.2; COD-load 6 (acetate), COD/N=52
16 De Kreuk and Van Loodrecht (2004)	1.3	14	16.5	EBPR, granule density: 78	SBR, 20 °C, anaerobic feeding (1h); $v_{p,min}$ 12; v_{gs} 2.5; DO 20%; COD-load 1.6 (acetate), COD/N=8
17. McSwain et al. (2004)	n.a.	46	9.0	filaments increased with length feast	SBR; v_{gs} 1.2; $v_{p,min}$ 15; COD-load 2.4, (glucose and peptone), COD/N=15.3
18. Schwarzenbeck et al. (2004)	n.a.	app. 30	6-7	51% particle removal (protozoa)	SBR; 17 °C; v_{gs} 0.5 – 0.7; COD-load 3.2 (barley dust), COD/N=38
19. Tay et al. (2001) and Tay et al. (2004)	flocs; 0.39; 0.37; 0.33	170; 62; 55; 46	1.4; 5.4; 6.5; 6.9	higher density and EPS with higher v_{gs}	SBR; 25 °C; $v_{p,min}$ 4.9; v_{gs} 0.3 (DO 1.5 mg l^{-1}), 1.2 (DO>3 mg l^{-1}), 2.4, 3.6; COD-load 6.0 (acetate), COD/N=10
20. Mosquera et al. (submitted)	flocs; 2	200; 48-75	0.9; 5.0	Disintegration at low DO. Granule density 13, 53	SBR (airlift); 20°C; DO 40%, 100% (v_{gs} 2.5); $v_{p,min}$ 12 ; load 1.6 (acetate), COD/N=8.4

AOB = ammonia oxidising bacteria; DO = Dissolved oxygen concentration in mg·l^{-1} or % of saturation concentration; EBPR = enhanced biological phosphate removal; n.a. = not available; app. = estimated by reading the graphs; BOD-, COD- and N-load in kg·m^{-3}·d^{-1}; granule and biomass density in g l^{-1}; v_{gs} = superficial gas velocity in cm·s^{-1}; $v_{p,min}$ = minimal settling velocity of the particles (as set by settling time) in m·h^{-1};

Table 2 Dependency of granule formation of some investigated parameters (+ = increased, - = decreased, no trend = no clear trend was observed when all results from table 1 were compared)

Process parameter	diameter	SVI	SS Concentration	density
Average value (standard deviation)	1.75 mm (1.1)	64 ml g^{-1} (29.8)	7 g l^{-1} (3.3)	44 g l^{-1} (22.0)
Type of substrate	glucose > molasse > ethanol = acetate > phenol	ethanol > glucose > acetate > phenol	no trend	acetate > glucose > ethanol[a]
Increased COD load	++	no trend	no trend	--
Increased C/N ratio	++	+	no trend	-
Settling time	no trend	no trend	no trend	no trend
Superficial gas velocity (shear)	maximum around 0.02 m s^{-1}	no trend	no trend	only measured at 0.025 m s^{-1}
Oxygen concentration	Not available, most oxygen concentrations reported as "larger than", but flocculation when aeration stops at low oxygen concentrations combined with pulse feed[5, 19, 20]			

[a]phenol not reported

A Comparative Study of Aerobic Granulation in Pilot- and Laboratory-Scale SBRs

Joo-Hwa Tay*, Qi-Shan Liu, Yu Liu, Kuan-Yeow Show, Volodymyr Ivanov and Stephen Tiong-Lee Tay

*Environmental Engineering Research Centre, School of Civil and Environmental Engineering, Nanyang Technological University, 50 Nanyang Avenue, Singapore 639798 (E-mail: cjhtay@ntu.edu.sg)

Abstract The development of aerobic granules in a pilot-scale sequencing batch reactor was investigated. For comparison, a laboratory-scale reactor was also operated in parallel. Both reactors were started up by seeding with 410 mg L^{-1} of pre-cultivated aerobic granules. The seed granules in the pilot-scale reactor disintegrated completely into loose flocs within the first 5 days of operation. However, aerobic granules were eventually re-formed from the floc particles in the reactor. On the contrary, the seed granules in the laboratory-scale reactor did not break up but remained relatively intact. There was a significant difference in the biomass distribution along the height of the two reactors. The biomass was evenly distributed along the height of the pilot-scale reactor, while more biomass was retained in the bottom half of the laboratory-scale reactor. The results pointed to the existence of different hydrodynamic conditions in the two reactors, which affected the biomass distributions within the two reactors and likely contributed to the disintegration of seed granules in the pilot-scale reactor. Compared to the granules cultivated in the laboratory-scale reactor, the granules developed in the pilot-scale reactor exhibited relatively larger granule sizes, weaker physical strengths, lower volatile solids contents and microbial activities.

Keywords Aerobic granules, granule formation, pilot-scale reactor, SBR

Introduction

Aerobic granulation is a recent innovation in biological wastewater treatment (Morgenroth et al., 1997; Peng et al., 1999; Beun et al., 2000; Tay et al., 2001). Granulation is a self-immobilization process in which biological solids agglomerate and develop into dense and compact granular biomass under controlled operating conditions. Similar to anaerobic granules, aerobic granules have a number of advantages over conventional biological flocs, such as regular and strong structure, good settling ability, high biomass retention, and ability to withstand high organic loadings (Beun et al., 1999; Tay et al., 2001; Moy et al., 2002).

Aerobic granules are highly versatile and have been successfully grown on a variety of substrates, such as glucose and acetate (Tay et al. 2002), molasses (Morgenroth et al. 1997), ethanol and phenol (Jiang et al., 2002; 2004). These aerobic granules can be cultivated over a wide range of organic loading rates ranging from 1.5 to 12.0 kg COD m^{-3} d^{-1} (Liu et al., 2003; Moy et al., 2002). The enhancement of nitrifying and phosphorus accumulating activities in aerobic granules has also been reported (Lin et al., 2003; Liu et al., 2004; Yang et al., 2003).

Aerobic granules can therefore be applied for organic and nutrient removal, as well as toxic wastewater treatment. However, it should be pointed out that the studies of aerobic granulation have thus far been confined to laboratory-scale reactors. Limited work has been reported to investigate the formation of aerobic granules in pilot-scale reactors, which is an important step for application of aerobic granulation in industrial-scale systems. The purpose of this study is to investigate aerobic granulation in a pilot-scale column-type sequencing batch reactor (SBR). In comparison, a smaller laboratory-scale reactor was also operated in parallel. This research is expected to provide useful information for scale-up of aerobic granulation technology.

Materials and methods

Experimental set-up

The pilot-scale column-type SBR had an internal diameter of 19 cm and a height of 160 cm. A working height of 120 cm was used in the study, giving a working volume of 34.0 L. Fig. 1 shows the schematic diagram of the pilot-scale reactor system setup. The reactor was operated at 4 hours per cycle with 12 min of influent filling, 221-224 min of aeration, 2-5 min of settling and 2 min of effluent withdraw. Effluent was discharged through a port 60 cm above the reactor bottom to achieve a volumetric exchange ratio of 50%.

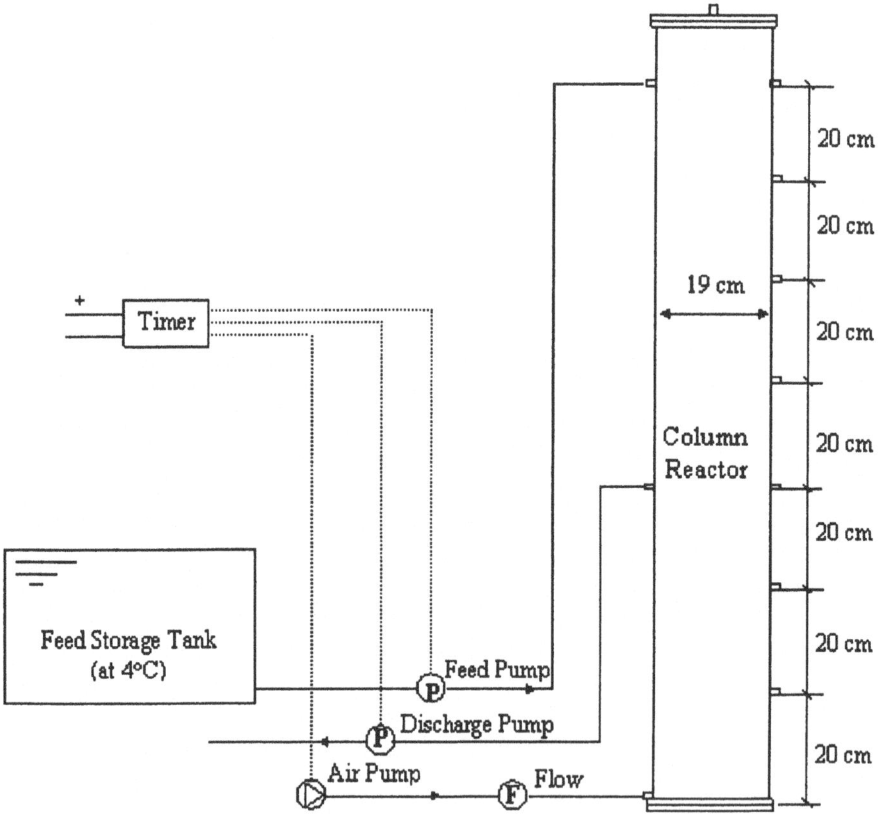

Figure 1: Schematic diagram of pilot-scale SBR setup

A ceramic disc diffuser (11 cm in diameter) was placed at the bottom of the column to supply air, and the air flow was controlled at 40 L min^{-1}, which was equivalent to 2.4 cm s^{-1} of superficial upflow air velocity. A laboratory-scale column reactor with the same reactor height as the pilot-scale reactor but a smaller diameter of 6 cm was operated in parallel for the purpose of comparison. The working volume of the laboratory-scale reactor was 3.4 L. Two spherical 3 cm diameter air stones that provided a similar bubble size as the disc diffuser were placed at the bottom of the column. The air flow was controlled at 4.0 L min^{-1}, giving the same superficial upflow air velocity as the pilot-scale reactor. The experiments were conducted in a temperature control room maintained at 25°C.

Medium and inoculum

A synthetic wastewater with sodium acetate as sole carbon source was used as described previously (Tay et al. 2001). A substrate COD concentration of 800 mg L^{-1}, equivalent to a loading rate of 2.4 kg COD m^{-3} d^{-1}, was applied to both reactors. The pilot-scale reactor was inoculated with 400 mL (1.2% of reactor volume) of aerobic granules cultivated with acetate in a small column reactor (5 cm diameter). The laboratory-scale reactor was inoculated with 40 mL of the same inoculum. This resulted in an initial biomass concentration of 410 mg L^{-1} in both reactors at start-up. The seed granules had a mean size of 0.83 mm and a sludge volume index (SVI) of 19 mL g^{-1}.

Analytical procedures

Effluent samples were analyzed for chemical oxygen demand (COD), and sludge samples were tested for mixed liquid suspended solid (MLSS), mixed liquid volatile suspended solid (MLVSS), sludge volume index (SVI), specific gravity, and specific oxygen uptake rate (SOUR) using standard methods (APHA, 1998). Granule size was measured by a laser particle size analysis system (Malvern MasterSizer Series 2600, Malvern Instrument, UK), or an image analysis (IA) system with an Olympus SZX9 microscope (Image-Pro Plus, V4.0, Media Cybernetics). Morphology of granules in terms of roundness and aspect ratio was analysed using an IA technique. Roundness of granule (0 = line, 1 = circle) was calculated as: 4π area/perimeter^2, in which perimeter was the length of the granule's outline. Aspect ratio was the ratio between minor axis and major axis of ellipse equivalent to the granule (0 = line, 1 = circle). Granule strength was determined according to the method of Ghangrekar et al. (1996) and expressed as an integrity coefficient (%) defined as the ratio of residual granules to the total weight of the granular sludge after 5 minutes of shaking at 200 rpm on a platform shaker.

Results

Observation of morphology change

Evolution of the morphology and structure of sludge in both reactors was tracked using the IA technique. The seed granules had a mean diameter of 0.83 mm (Fig. 2a). The granules in the pilot-scale reactor disintegrated shortly after start-up, and loose flocs dominated the reactor on day 5 (Fig. 2b). The mean diameter of the sludge particles decreased to 0.19 mm and SVI increased from 19 mL g^{-1} to 175 mL g^{-1} (Fig. 3a). These are typical values for conventional sludge flocs. The flocs gradually re-formed into compact aggregates (Fig. 2c), which on day 20 had a mean diameter of 0.4 mm and an SVI of 63 mL g^{-1}.

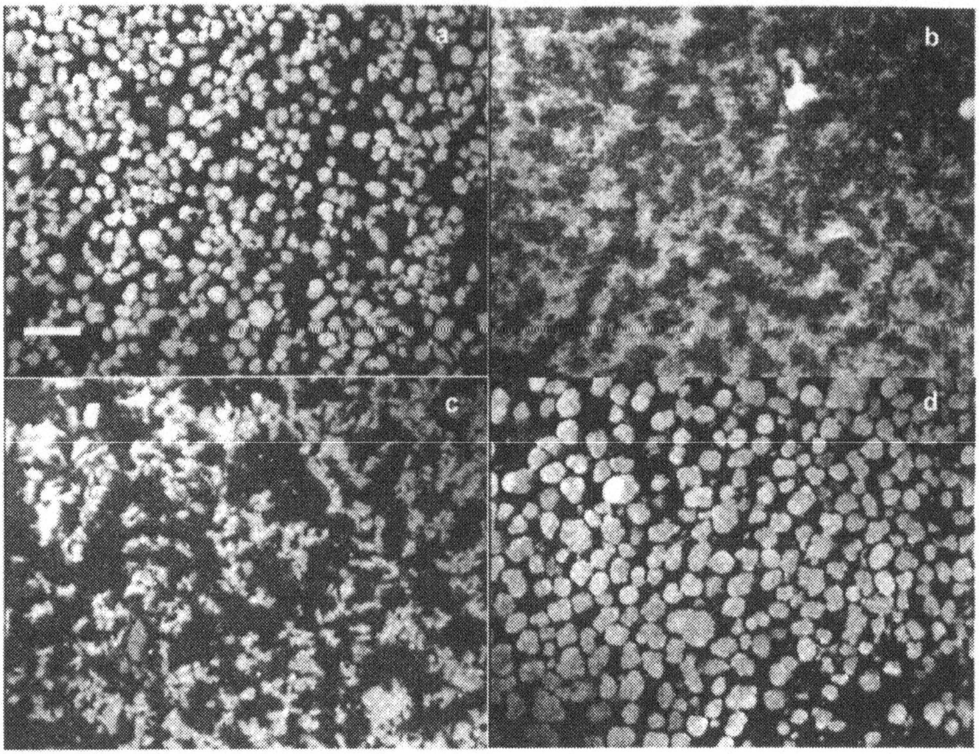

Figure 2: Sludge morphology in pilot-scale reactor at day 1 (a), day 5 (b), day 20 (c) and day 65 (d). Scale bar represents 4 mm.

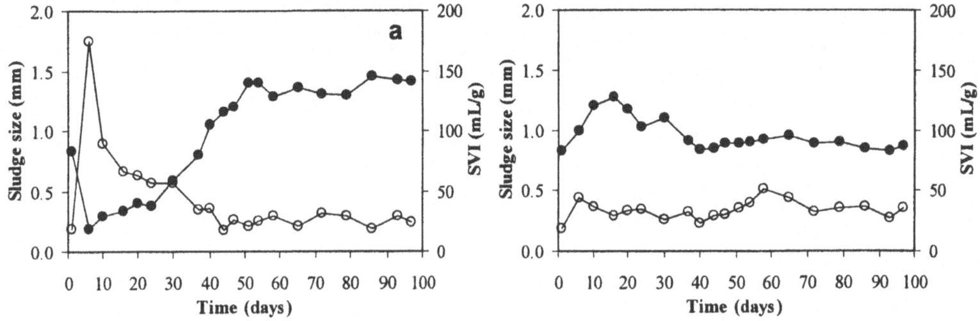

Figure 3: Mean diameter (•) and SVI (○) versus time in pilot-scale (a) and laboratory-scale (b) reactors

The granules increased in size to peak at a diameter of 1.4 mm on day 50 and finally stabilized at this level. The steady-state granules had a compact structure that was similar to the seed granules, but a larger size (Fig. 2a and 2d). The SVI stabilized at around 26 mL g^{-1} (Fig. 3a). Unlike the granules in the pilot-scale reactor, the seed granules in the laboratory-scale reactor did not break up but managed to maintain their structure. The mean granule diameter increased from 0.83 to 1.3 mm during the first 3 weeks, then decreased to stabilize at 0.9 mm (Fig. 3b). The SVI of the granular sludge in the laboratory-scale reactor remained relatively constant at about 34 mL g^{-1}.

Sludge growth and biomass distribution

The sludge concentration in the pilot-scale reactor was 0.4 g L^{-1} at the beginning of start-up (Fig. 4a) and gradually increased to 6.5 g L^{-1} within the first 3 weeks of operation. There was a drop in sludge concentration from day 30 to 40, which was attributed to a reduction in settling time from 5 to 2 min. The reduction in settling time caused slow settling sludge to wash out and resulted in a temporary reduction in sludge concentration. However, the shorter settling time also exerted a selection pressure to encourage the retention of sludge with better settleability. Consequently, a recovery in sludge concentration was observed, which finally stabilized at 8 g L^{-1}. The profile of sludge concentration with time followed a similar trend in the laboratory-scale reactor, and a comparable sludge concentration was maintained in the reactor at steady state.

The variation in sludge concentration along the reactor height was also determined. During the initial period, the sludge was distributed rather evenly along the reactor height in the pilot-scale reactor (Fig.5a), while more sludge was found in the lower half of the laboratory-scale reactor. Similar sludge distribution profiles were observed in both reactors towards the end of the operation when conditions had stabilized and the sludge concentration averaged 8 g L^{-1} (Fig. 5b). The differences in the sludge distribution profiles suggested that the hydrodynamic conditions were different in the two reactors.

Physical and microbial characteristics of granules

Granule size and morphology

The mean diameter of aerobic granules developed in the pilot-scale reactor was 1.37 mm (Table 1), which was larger than that of the seed granules of 0.83 mm. Granules cultivated in the laboratory-scale reactor had a similar size as the seed granules. Granules developed in the pilot-scale reactor and the laboratory-scale reactor ranged from 0.4 to 2.7 mm in diameter and 0.3 to 2.0 mm in diameter, respectively (Fig. 6). Both sets of granules exhibited a similar morphology in terms of aspect ratio and roundness (Table 1).

Settling property

Aerobic granules in the pilot-scale reactor had a low SVI of 26.5 mL g^{-1} and a high specific gravity of 1.017, as compared to those cultivated in the laboratory-scale reactor, which had an SVI of 34.4 mL g^{-1} and a specific gravity of 1.015. The volatile solids content of granules in the pilot-scale reactor (62.4%) was lower than that of granules cultivated in the laboratory-scale reactor (74.9%). This indicated a higher accumulation of inorganic compounds in the larger granules of the pilot-scale reactor. The higher inorganic content also corresponded with a lower SVI and could be a reason for the better settling characteristics of the granules in the pilot-scale reactor.

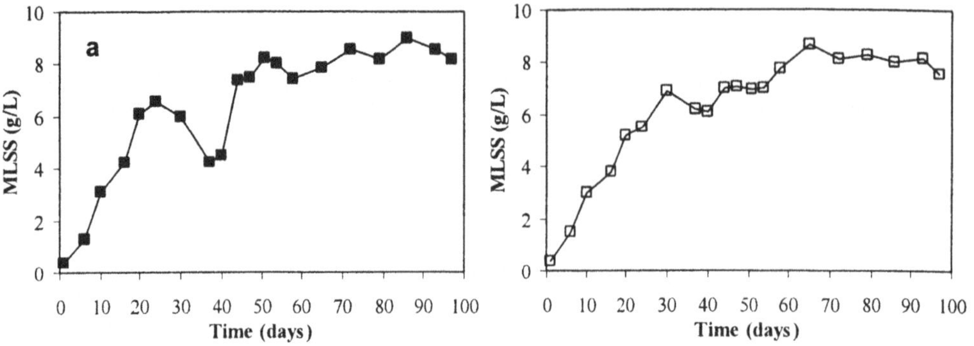

Figure 4: Sludge concentration versus time in pilot-scale (a) and laboratory-scale (b) reactors

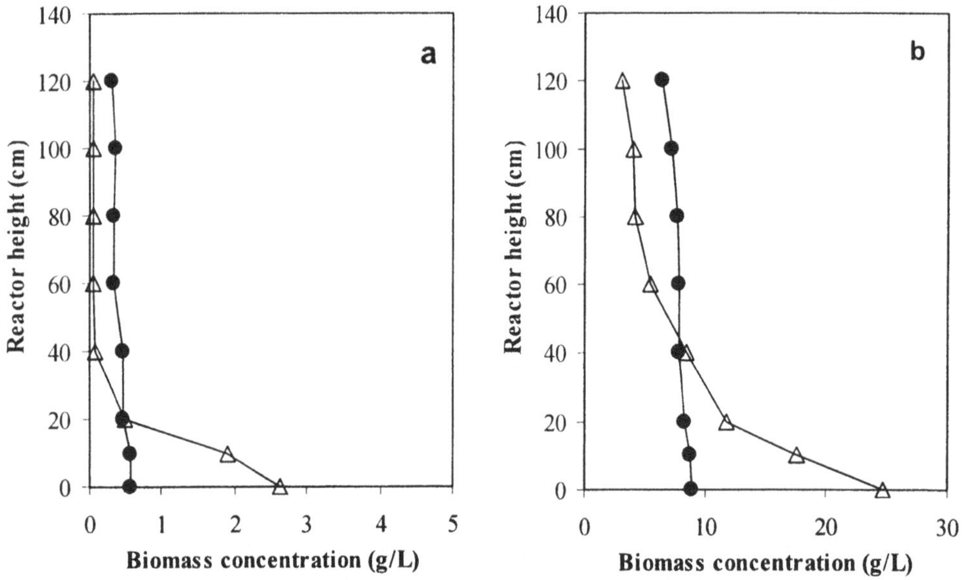

Figure 5: Sludge distribution along the reactor height during the first day (a) and at steady state (b). •: pilot-scale reactor; Δ: laboratory-scale reactor

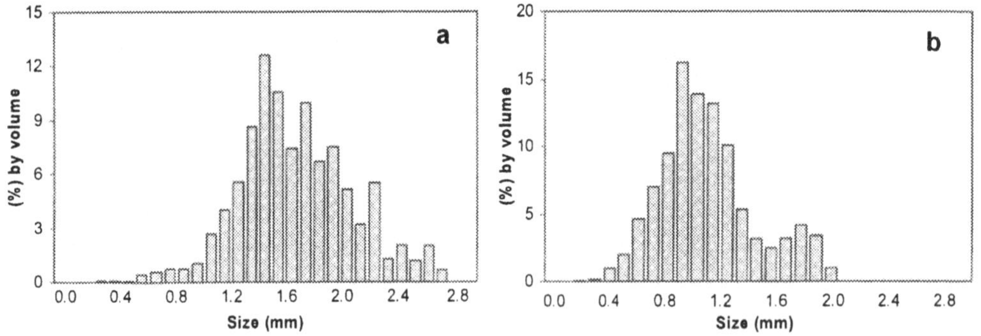

Figure 6: Comparison of size distribution of biomass in pilot-scale (a) and laboratory-scale (b) reactors at steady state

Physical strength

The physical strength of granules, expressed as integrity coefficient, was 96.0% for the pilot-scale reactor and 96.9% for the laboratory-scale reactor. The larger the integrity coefficient, the higher is the strength of granules. The results showed that the granules developed in the laboratory-scale reactor were marginally stronger than those in the pilot-scale reactor.

Microbial activity

The SOUR as a measure of microbial activity was 74.1 mgO_2 $gVSS^{-1}$ h^{-1} for the pilot-scale reactor and 80.6 mgO_2 $gVSS^{-1}$ h^{-1} for the laboratory-scale reactor (Table 1). The slightly lower microbial activity of granules from the pilot-scale reactor is thought to be size-related, because the limitation of mass transport and diffusion is generally more pronounced with larger granules.

Table 1: Characteristics of aerobic granules cultivated in pilot-scale and laboratory-scale SBRs

Items	Pilot-scale reactor	Laboratory-scale reactor
Mean diameter (mm)	1.37 (±0.09)	0.89 (±0.07)
Aspect ratio	0.67 (±0.16)	0.69 (±0.15)
Roundness	0.69 (±0.15)	0.69 (±0.16)
SVI (mL g^{-1})	26.5 (±5.9)	34.4 (±6.9)
Specific gravity	1.017 (±0.0005)	1.015 (±0.0005)
VSS/SS (%)	62.4 (±2.8)	74.9 (±3.6)
Integrity coefficient (%)	96.0 (±2.0)	96.9 (±2.5)
SOUR (mgO_2 $gVSS^{-1}$ h^{-1})	74.1 (±12.4)	80.6 (±18.2)

Discussion

Two SBRs were started up by seeding with 410 mg L^{-1} of pre-cultivated aerobic granules. The seed granules in the pilot-scale reactor disintegrated completely into loose flocs within the first 5 days of operation. However, aerobic granules were eventually re-formed from the floc particles in the reactor. On the contrary, the seed granules in the laboratory-scale reactor did not break up but remained relatively intact. The biomass concentration in both reactors increased gradually to stabilize at 8 g L^{-1} at steady state. We hypothesize that the disintegration of aerobic granules in the pilot-scale reactor is linked to the prevailing hydrodynamic conditions. As depicted in Fig. 5, different biomass distributions along the reactor height were observed in the two reactors. The biomass was evenly distributed along the reactor height of the pilot-scale reactor, while most of the biomass was concentrated in the bottom half of the laboratory-scale reactor. This indicated that the hydrodynamic characteristics could be different in the two reactors. It had been reported that the strength of the liquid circulations would increase with the increase in the scale of column (Krishna, et al., 2001). The strong circulations could accelerate the bubbles traveling upwards in the central core of the column, and subsequently strong upflow of the biomass. This might be the reason that a relatively even biomass distribution was observed along the reactor height in the pilot-scale SBR. With the larger diameter of the pilot-scale reactor, there would also be a lower wall effect. Moreover, the size and placement of air diffusers in the column would certainly affect the hydrodynamic pattern. In the study, a 11 cm disc diffuser was placed at the bottom of the pilot-scale reactor, while two 3.0 cm diameter spherical air stones were used for the laboratory-scale reactor. These diffusers occupied 34% and 50% of the cross-sectional area of the pilot-scale and laboratory-scale reactors, respectively. The different aerators should give rise to different hydrodynamic flow patterns in the two reactors, namely airlift mixing due to upflow of sludge in centre and downflow of sludge near the wall of the pilot-scale reactor, compared to a more even distribution of airflow across the entire cross-section of the laboratory-scale reactor. In fact, it is known that the shear stress is associated with the specific energy input which is related to the upflow air velocity (Mahnke et al., 2000), while the gas distribution system may also influence the shear stress in the column reactor. Thus, the shear stress may differ in these two reactors. Some of the seed granules that were originally cultivated in a small column reactor were transplanted into the pilot-scale reactor. This was akin to a shock episode that exposed the granules to a new set of environmental and hydrodynamic conditions, such as the high air and liquid velocities encountered in the central core of the pilot-scale reactor, and the seed granules disintegrated shortly after start-up. However, the repetitive nature of the SBR operation provided a selective process that allowed for the gradual redevelopment of granular biomass with good settling characteristics. The short settling and discharge times played a key role in setting up the selection pressures necessary for granule development. For instance, a short settling time forces slow settling bioflocs to wash out, and promotes the selective retention and subsequent proliferation of fast settling sludge aggregates (Qin et al., 2004).

An initial increase in granule size was observed in the laboratory-scale reactor in the first 2 weeks of operation. This could be due to the relatively low concentration of granular sludge in the reactor when the experiment started (Fig. 3b and 4b), which in turn led to less collisions among the granule particles and less detachment of biosolids from individual granules. The granule size decreased again when the biomass concentration increased in subsequent operation, and stabilized when biomass concentration stabilized. In fact, Tay et al (2004) showed that the size of the granules that developed was inversely correlated to the amount of shear generated from the rising air bubbles and collisions among the sludge particles. Granules in the laboratory-scale reactor were smaller than in the pilot-scale reactor, but similar in size to the seed granules (Fig. 3 and Table 1). The similarity in size of the seed granules and the granules developed in the laboratory-scale reactor should not be unexpected, since the reactor configuration and operating conditions in the laboratory-scale reactor and the reactor used for pre-cultivation of seed granules were similar. As discussed earlier, the wall effect would appear to be more significant in the laboratory-scale reactor, which should result in a greater frequency of collisions among the granules and a higher shear. Moreover, as exhibited in Fig 5, as high as 25 g L^{-1} of granular sludge was retained in the bottom half of the laboratory-scale reactor, whereas 8 g L^{-1} of granular sludge was evenly distributed in the pilot-scale reactor. The high accumulation of granular sludge at the bottom of the laboratory-scale reactor would certainly increase the collision and detachment rate among the granule particles. Therefore, smaller granules were developed in the laboratory-scale reactor.

Conclusions

The development of aerobic granules in a pilot-scale SBR seeded with aerobic granules pre-cultivated in a small column reactor was investigated. It was observed that the seed granules disintegrated in the first a few days of operation, but the disintegrated biomass would re-form stable granules latterly. On the contrary, the seed granules could be successfully maintained in the laboratory-scale reactor without significant disintegration. The different hydrodynamic patterns encountered in the pilot- and lab-scale reactors might be the reason for the observed phenomena. Factors such as reactor diameter and wall effect, as well as size and placement of air diffusers, would influence the hydrodynamic conditions in the reactors, which in turn determine the properties of the granules.

References

APHA, (1998). *Standard Methods for the Examination of Water and Wastewater*. 19th ed. American Public Health Association, Washington, DC, USA.

Beun, J.J., Hendriksm A., van Loosdrecht, M.C.M., Morgenroth, E., Wilderer, P.A. and Heijnen, J.J. (1999). Aerobic granulation in a sequencing batch reactor. *Wat. Res.* **33**(10), 2283-2290.

Beun, J.J., van Loosdrecht, M.C.M., Morgenroth, E. and Heijnen, J.J. (2000). Aerobic granulation. *Wat. Sci. Technol.* **41**(4-5), 41-48.

Ghangrekar, M.M., Asolekar, S.R., Ranganathan, K.R. and Joshi, S.G. (1996). Experience with UASB reactor start-up under different operating conditions. *Wat. Sci. Technol.* **34**(5-6), 421- 428.

Jiang, H.L., Tay, J.H. and Tay, S.T.L. (2002). Aggregation of immobilized activated sludge cells into aerobically grown microbial granules for the aerobic biodegradation of phenol. *Lett. Appl. Microbiol.* **35**(5), 439-445.

Jiang, H.L., Tay, J.H. and Tay, S.T.L. (2004). Changes in structure, activity and metabolism of aerobic granules as a microbial response to high phenol loading. *Appl. Microbiol. Biotechnol.* **63**, 602 – 608.

Krishna R., van Baten J. M., Urseanu M. I. (2001) Scale effects on the hydrodynamics of bubble columns operating in the homogeneous flow regime. *Chem. Eng. & Tech.* **24** (5), 451-458.

Lin, Y.M., Liu, Y. and Tay, J.H. (2003). Development and characteristics of phosphorous-accumulating microbial granules in sequencing batch reactors. *Appl. Microbiol. Biotechnol.* **62**, 430–435.

Liu Q.S., Tay, J.H. and Liu, Y. (2003). Substrate concentration-independent aerobic granulation in sequential aerobic sludge blanket reactors. *Environ. Technol.* **24**, 1235-1242.

Liu, Y. and Tay, J.H. (2002). The essential role of hydrodynamic shear force in the formation of biofilm and granular sludge. *Wat. Res.* **36**, 1653–1665.

Liu, Y., Yang, S.F. and Tay, J.H. (2004). Improved stability of aerobic granules by selecting slow-growing nitrifying bacteria. *J. Biotechnol.* **108**, 161-169.

Mahnke E. U., Büscher K., Hempel D. C. (2000) A novel approach for the determination of mechanical stresses in gas-liquid reactors. *Chem. Eng. & Tech.* **23** (6), 509-513.

Morgenroth, E., Sherden, T., van Loosdrecht, M.C.M., Heijnen, J.J and Wilderer, P.A. (1997). Aerobic granular sludge in a sequencing batch reactor. *Wat. Res.* **31**, 3191-3194.

Moy, B.Y.-P., Tay, J.H., Toh, S.K., Liu, Y. and Tay, S.T.L. (2002). High organic loading influences the physical characteristics of aerobic granules. *Lett. Appl. Microbiol.* **34**(6), 407-412.

Peng, D., Bernet, N., Delgenes, J.P. and Moletta, R. (1999). Aerobic granular sludge-a case study. *Wat. Res.* **33**, 890-893.

Qin, L., Tay, J.H. and Liu Y. (2004). Selection pressure is a driving force of aerobic granulation in sequencing batch reactors, *Proc. Biochem.* **39**, 579-584.

Tay, J.H., Liu, Q.S. and Liu, Y. (2001). Microscopic observation of aerobic granulation in sequential aerobic sludge blanket reactor. *J. Appl. Microbiol.* **91**, 168-175.

Tay, J.H., Liu, Q.S. and Liu, Y. (2002). Characteristics of aerobic granules grown on glucose and acetate in sequential aerobic sludge blanket reactors. *Environ. Technol.* **23**, 931–936.

Tay, J.H., Liu, Q.S. and Liu, Y. (2004). The effect of upflow air velocity on the structure of aerobic granules cultivated in sequencing batch reactor. *Wat. Sci. Technol.* **49**(11-12), 30-34.

Yang S.F., Tay, J.H., Liu, H., (2003). A novel granular sludge sequencing batch reactor for removal of organic and nitrogen from wastewater, *J. Biotechnol.* **106**, 77-86.

Promising results pilot research aerobic granular sludge technology at WWTP Ede

L.M.M. de Bruin*, H.F. van der Roest*, M.K. de Kreuk, M.C.M. van Loosdrecht****

* DHV Water B.V., P.O. Box 484, 3800 AL Amersfoort, The Netherlands (e-mail: bart.debruin@dhv.nl)
** Kluyverlaboratory for Biotechnology, Delft University of technology, Julianalaan 67, 2628 BC Delft, The Netherlands (e-mail: m.dekreuk@tnw.tudelft.nl)

Abstract A large pilot research project was started in order to demonstrate the applicability of the aerobic granular sludge technology for the treatment of municipal wastewater under Dutch circumstances. Several operational philosophies were tested to learn at which conditions granulation occurs with municipal wastewater as a substrate. Fast formation of granules was observed under conditions of extensive COD-removal, extensive biological phosphate removal and low nitrate effluent concentrations. The potential of the technology is very promising since complete granulation with municipal wastewater as substrate was shown and extensive nutrient removal seems well feasible.

Keywords aerobic granular sludge, municipal wastewater, nutrient removal, pilot plant research

Introduction

Since 1999 DHV Water BV and the Delft University of Technology (TUD) have been cooperating closely on the development of the aerobic granular sludge technology. In the period between 1999 and 2002 an extensive laboratory research and a feasibility study were carried out. Based on the good results a pilot research was started up in September 2003 and this article describes the results up to now. The results of the laboratory tests including the selection mechanism and the feasibility study have been published in several papers and are summarised briefly below.

Selection mechanism for granulation

Although the exact selection mechanism for granulation is not known yet, the following process conditions are assumed to play an important role: 1) a high hydraulic surface load during sedimentation resulting in the selection of well settleable biomass, 2) initial high substrate concentrations during substrate conversion, as in batchwise operation, 3) conversion of easily degradable substrate into slowly degradable storage products and 4) sufficient shear forces leading to the formation of smooth and dense granules.

Laboratory research

During recent laboratory research stable granulation was obtained in a batchwise operated reactor with a three-minute pulse feeding period (total cycle time 3 hours), feeding of synthetic wastewater consisting of acetate and ammonium as C- and N-source respectively, and a high dissolved oxygen (DO) concentration during the aerated period (Beun et al. 2001). At low DO (40%) and a longer feeding period the granules became unstable and fell apart. Research had only been carried out in air-lift reactors at that point.

For a full-scale application it is essential to allow longer feeding periods and operation at low DO. Next to this, for economical reasons a bubble column is preferred over an airlift reactor. Further laboratory research aimed at elimination of these bottlenecks for scale-up and at evaluation of the economical and technical feasibility of the granular sludge sequencing batch reactor (GSBR).

Laboratory research focussed mainly on the possibilities for simultaneous nutrient removal. Therefore, the feeding period was varied and granulation at low DO was studied. The performance of a bubble column compared to an air-lift reactor was determined as well.

In order to provoke a "pulse feeding" it was proposed to feed during 33 % of the cycle time under anaerobic conditions. If substrate is not converted during this period, substrate concentrations would be similar to a pulse feeding operation when aeration starts. Moreover, feeding influent in an upflow mode through the granular sludge blanket will allow bacteria involved in biological phosphate removal to proliferate. In this case, substrate is taken-up by PAO cells during the anaerobic feeding period and is cell-internally stored as a polymer. It was shown that a 60 minutes feeding period with synthetic wastewater (total cycle time 3 hours) and a DO of 20% resulted in complete COD and extensive nutrient removal (de Kreuk et al. 2004). Biomass concentrations that can be maintained in this type of reactors were around 5 times higher than in an activated sludge system. Except for a longer start-up period the performance in a bubble column was similar compared with an air-lift reactor. Finally, formation of aerobic granular sludge with pre-treated sewage was shown. These results clearly showed the potential of the technology.

Feasibility study

The feasibility study showed that the aerobic granular sludge technology seems very promising (de Bruin et al. 2004). Based on total annual costs a GSBR with pre-treatment and a GSBR with post-treatment prove to be more attractive than the reference activated sludge alternatives (6-16%). A sensitivity analysis shows that the GSBR technology is less sensitive to land price and more sensitive to Rain Water Flow (RWF). In the Netherlands, average values for the ratio of RWF and DWF (Dry Weather Flow) amount to 3-4. The GSBR technology becomes even more attractive at lower RWF/DWF ratios and higher land prices. Because of the high allowable volumetric load the footprint of the GSBR variants is only 25% compared to the references. However, the GSBR with only primary treatment cannot meet the present effluent standards for municipal wastewater in the Netherlands, mainly because of exceeding the suspended solids effluent standard caused by washout of not well settleable biomass.

Moreover, municipal wastewater treatment plants (WWTPs) in the Netherlands are going to be faced with more stringent effluent standards. In general, activated sludge plants will have to be extended with a post treatment step (e.g. sand filtration) or be transformed into a Membrane Bio Reactor. In this case a GSBR variant with primary treatment as well as post treatment can be attractive alternatives.

Pilot Research

Based on both the good laboratory research results and the positive technical and financial evaluation for full-scale application, a large pilot research project was started in September 2003. The pilot plant is located at the WWTP Ede. Main objective is to demonstrate the applicability of the aerobic granular sludge technology for the treatment of municipal wastewater under Dutch conditions (e.g. high RWF/DWF ratios). In order to achieve this objective, the following questions have to be answered:

- is a stable granulation process under field conditions possible?
- can the present (and future) effluent standards in the Netherlands be met? If possible, which are the required process conditions?
- what are the properties of the granules and any fines which are washed out with the effluent?

Description Pilot Plant

The pilot plant is mainly placed in a container. Figures 1 and 2 show a process flow diagram and a photo of the pilot plant. The hydraulic design capacity of the pilot plant amounts to 5 m3/h. The heart of the installation consists of two parallel biological reactors with heights and diameters of 6 m and 0,6 m respectively. Both reactors can be operated as an airlift reactor or as a bubble column. Aeration takes place by means of a blower in combination with a diffuser and the airflow is automatically controlled based on continuous DO measurements. The effluent can be withdrawn at a height of 4 m by means of a automatically controlled valve or at a height of 6 m via a weir. After the GSBRs two tanks are located that serve as buffers in order to check the effluent quality properly. The pre-treatment of the pilot plant consists of a primary sedimentation tank, optionally followed by a pressurised sand filter. If desired, polymer can be dosed in order to improve the efficiency of the primary sedimentation tank.

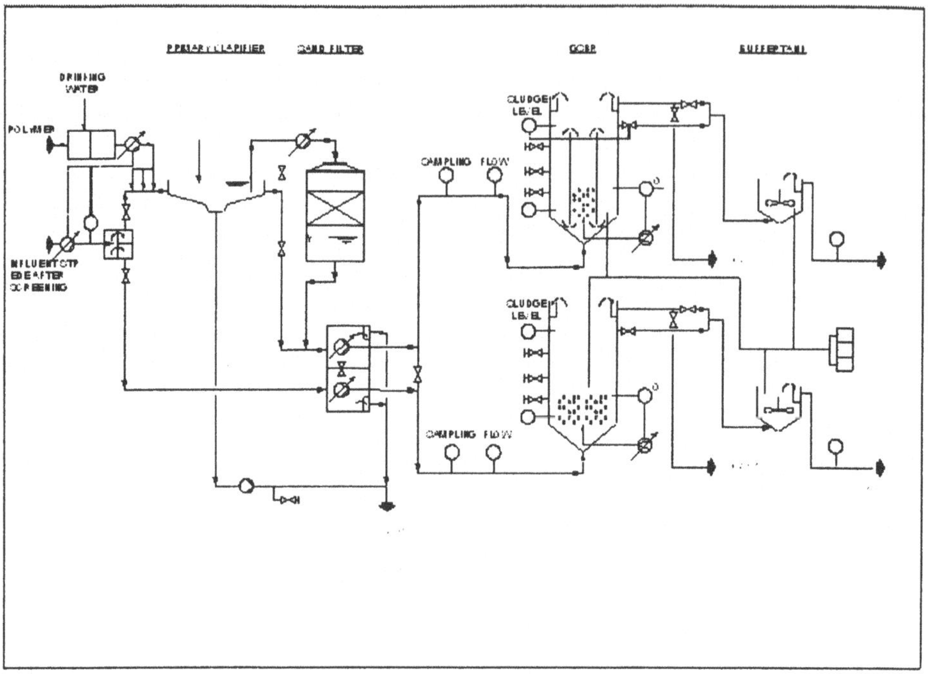

Figure 1: Process Flow Diagram pilot plant

Figure 2: Photograph pilot plant

The pre-treatment can be bypassed and in this case the GSBRs can be fed with raw, screened influent. Influent can be fed at a constant feed, but it is also possible to control the influent flow similar to the influent pattern of the STP Ede. Special feature of the pilot plant is a high degree of automation that is partly based on online monitors such as oxygen, ammonium, nitrate, phosphate. In case of process failures or disturbances, process operation is changed automatically. Moreover, remote process monitoring and operation is possible.

Preliminary Results and Evaluation

The pilot plant was started up in October 2003 and up to now it has been operated at a constant influent flow rate. The average influent composition was as follows: 560 mg/l COD, 225 mg/l SS, 58.4 mg/l N-Kj, 10.0 mg/l Ptotal. One of the reactors (GSBR 1) has always been operated as a bubble column. The second reactor had been operated as an air-lift reactor (GSBR 2) from October 2003 until May 2004. Because the performance of the bubble column was at least comparable with the air-lift reactor, the last mentioned reactor was rebuilt into a bubble column in May 2004. This was an somewhat unexpected result because an air-lift reactor was expected to perform better (higher shear forces) compared to a bubble column. Since May 2004 the only difference between the two reactors is the DO during aeration (2 mg/l for GSBR 1 and 5 mg/l for GSBR 2).

The cycle time has been varied between 2,5 and 4 hours. The anaerobic feed time amounts to 50-60 minutes, the hydraulic surface load during sedimentation amounts to 3-4 m/h and the decantation time is 20-30 minutes. The remaining time consists of aeration.

In the period between October 2003 and July 2004 the pre-treatment consisted of sedimentation without dosing of chemicals, resulting in an average removal efficiency of 55% for SS, 30% for total COD, 9% for TKN and 15% for Ptotal. In July 2004 the pre-treatment was extended with a pressurized sand filter, aiming at a higher removal efficiency for SS and stimulating granulation.

Table 1: Key data on operation, sludge characteristics and effluent quality.

Parameter	Unit	GSBR 1				GSBR 2			
		17/2-12/3	12/3-26/3	26/3-28/6	2/7-date	17/2-12/3	12/3-26/3	26/3-28/6	2/7-date
Sludge load & characteristics									
COD-load	$kg/(m^3.d)$	1,0	0,5	1,7	2,0	1,0	0,5	1,7	2,1
DS	kg/m^3	2,1	2	3	4	2	1	2	3
SVI 30	ml/g	221	170	104	84	173	163	126	97
Temperature	°C	14	18	17	21	14	18	17	21
Effluent quality									
COD	mg/l	171	110	119	99	195	148	132	100
COD filtered	mg/l	55	41	56	52	58	42	54	51
NH_4	mg/l	4,3	2,0	18,4	16,5	10,3	0,2	14,9	5,6
NO_3	mg/l	13,0	14,3	3,8	2,4	13,0	14,3	3,8	2,4
P-ortho	mg/l	2,8	3,9	0,9	0,6	2,3	4,3	0,5	0,8
SS	mg/l	81	39	75	63	106	69	85	62

The pilot was inoculated with activated sludge from the WWTP Ede in October 2003 and at that point automatic aeration control was not possible. Initially, granulation seemed to make good progress resulting in good sludge settling properties. However, due to low process temperatures down to 5 °C and inadequate aeration control leading to high nitrate concentrations and poor biological phosphate removal, granulation did not proceed. Therefore, in January 2004 the pilot plant was adjusted by implementation of aeration control based on DO measurements. In addition the walls of GSBRs were insulated and also the influent flow could be heated artificially in order to control the (minimum) process temperature.

In February 2004 the installation was started up again. The operational philosophy was to achieve high removal efficiencies with regard to biological phosphate removal and nitrification. At this time biological phosphate removal was on the average 65% and nitrification was not complete. Because ammonium concentrations in the effluent were rather high and the sludge settling characteristics seemed to deteriorate, the COD-load was reduced with 50% at 12 March. This change of operation resulted in full nitrification and higher nitrate concentrations in the effluent. At the same time biological phosphate removal deteriorated dramatically and granulation did not improve.

The aforementioned results again necessitated a change of the operational philosophy. During start-up, it proved to be impossible to maintain full biological phosphate removal and nitrification at the same time. Since the selection of slow growing organisms (PAOs in this case) was expected to be a requirement for granulation, the COD-load was increased at the end of March and from then on granulation seemed to start. An indication for this was improved settling properties (SVI30), but there was a need for more information on granulation. Because of this, granules size distribution has been determined at a weekly basis since the end of April 2004. All sludge particles with a diameter higher than 212 μm are defined as granules.

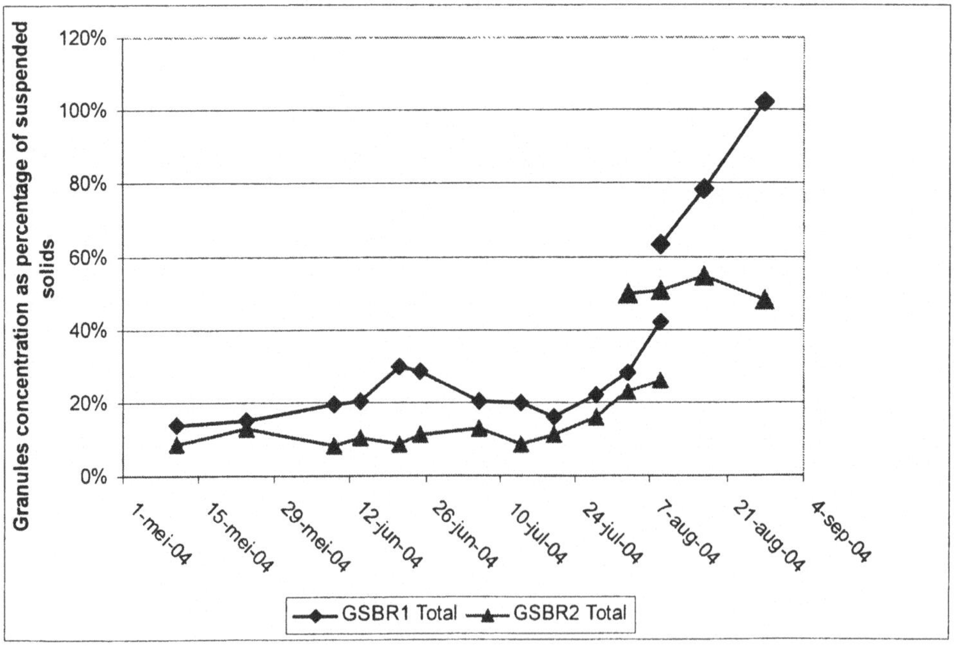

Figure 3: Development of granulation

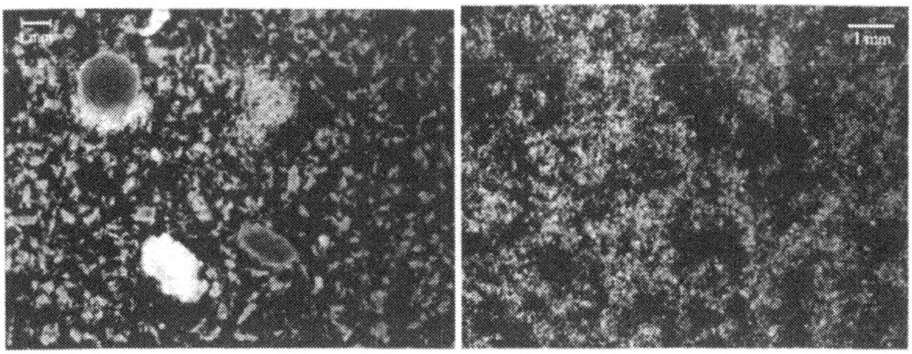

Figure 4: Photographs of sludges Granules from GSBR 1 (65% granulation, left); Sludge from WWTP Ede (right)

Granulation seemed to proceed fairly well and at the end of June around 30% of the sludge in GSBR 1 consisted of granules. Unfortunately, because of an operational failure more than 75% of the sludge was lost in both reactors at the end of June. Again, the reactors were inoculated with sludge from the WWTP Ede. From this time on granulation went quickly. This is illustrated in the figures 3 and 4. The leap in granulation in early August was caused by a change of the sludge sampling method. With the previous sampling method a large part of the granules was crushed. At the end of August, granulation in GSBR 1 was complete, meaning all sludge parts were larger than 212 μm. The SVI 30 was at that point 55 ml/g in GSBR 1. The size of the granules is expected to increase in the period to come. Granulation in GSBR 2 proceeds slower and was around 50% at the end of August with a SVI 30 of around 70 ml/g.

As the granulation process moved forward, sludge wash out decreased and as a result the biomass concentrations in the reactors increased. In the present operation mode, sludge wash out is to be expected because surplus sludge is not separately discharged from the reactors.

Although the current operational philosophy is to achieve high ammonium effluent concentrations, nitrification occurs for the greater part, especially in rainy periods with low influent concentrations. However, it seems that at a certain granulation level nitrification can be allowed because nitrate is converted simultaneously in the core of the granules.

Conclusions and Follow-Up

Because of various reasons the start-up period required more time than anticipated and the start-up has not been completed fully yet. Granulation is expected to continue, resulting in larger granules.

Many lessons have been learnt and it is believed that one of the key factors for granulation is the conversion of easily degradable COD into storage products that are converted consequently at a slower rate. In the process of municipal wastewater treatment this is achieved through the route of biological phosphate removal.

Shear forces do not seem to be a major factor since air-lift reactor and bubble column showed almost the same performance. In addition, the advantage of higher shear forces in GSBR 2 due to a higher DO does not outweigh disadvantages such as higher nitrate concentrations in the effluent. The hydraulic surface load during sedimentation seems to play a minor role in the granulation process. Good settling properties are more a result of the formation of granules.

Although the research is more or less in the initial stage and many questions are still to be answered, the potential of the technology is very promising since complete granulation with municipal wastewater as substrate was shown and extensive nutrient removal seems well feasible.

Acknowledgements

This research is funded by the Dutch Foundation for Water Research (STOWA) as well as by the Dutch technology Foundation (STW).

References

Beun J.J. Van Loosdrecht M.C.M. and Heijnen J.J. (2001). N-removal in a Granular Sludge Sequencing Batch Airlift Reactor. Biotechnol. Bioeng. 75(1), 82-92.

De Bruin, L.M.M., De Kreuk, M.K., van der Roest, H.F., Van Loosdrecht M.C.M. and Uijterlinde C.A. (2004). Aerobic Granular Sludge Technology, alternative for Activated Sludge technology? Wat. Sci. Technol. 49(11-12), 1-9.

De Kreuk, M.K. and Van Loosdrecht M.C.M. (2004). Selection of Slow Growing Organisms as a means for Improving Aerobic Granular Sludge Stability. Wat. Sci. Technol. 49(11-12), 9-19.

Mosquera-Corral, A., De Kreuk M.K., Heijnen J.J. and Van Loosdrecht M.C.M. (Submitted). Effects of Dissolved Oxygen on N-removal in an Aerobic Granular Sludge Reactor. Wat. Res.

Risks of aerobic granular sludge technology; ethical and methodological aspects

I.R. van de Poel*, S.D. Zwart*†, M. Brumsen*, H.G.J. van Mil*

* Delft University of Technology, School of Technology, Policy and Management, Section Philosophy, P.O. Box 5015, 2600 GA Delft, The Netherlands.
† Eindhoven Technical University, Department of Technology Management, Section Philosophy and Ethics of Technology, PO Box 513, 5600 MB Eindhoven, The Netherlands.

Abstract

In this paper we report on the first results of our STW project, which considers ethical and methodological aspects of the research into the design and development of GSBRs in the Netherlands. To get a first idea about the risks of this development, we have invited ten specialists from the water boards, universities, engineering firms and regulators and asked them about their ideas regarding the risks involved. Our question as to what those risks are resulted in a shortlist of seventeen risks; then we asked them to score those risks on ordinal probability and impact scales, This resulted in a distinction between large and small risk and a number of medium scale risks. Finally we asked the participants to indicate in what phase of the GSBR development the various risks should be addressed. For some risks this showed that if undesirable consequences should occur there would be a problem of many hands. We will present an overview of the scores and provide some interpretations.

Keywords Granular sludge, risks, risk perceptions, ethics, methodology, GDR, problem of many hands.

Introduction

In January 2004, we started research parallel to the "Aerobic Granular Sludge Reactors" (GSBR) project under supervision of Prof. van Loosdrecht at the Kluyver Laboratory of the Technical University of Delft on which De Bruin et al. report in their paper in the present volume. Our parallel research concerns the ethical and methodological aspects of risks involved in this new sewage treatment technology in general, and more in particular the transition from a lab-scale model to a full-size sewage water treatment plant.

Almost all technological developments have considerable effects on the well-being of fellow citizens or on that of the environment and its inhabitants. Not only do technological developments have an impact on people and the environment, in addition, in many instances the knowledge about a new technology is only partial. This lack of knowledge not only relates to mechanisms known to be involved but also to the possibility of processes that remain completely unobserved. The effects to people and the environment, and the partiality of knowledge jointly imply that new technological developments often pose *potential* risks to people or their environment. As a result, the environment or people, often without their consent, could be harmed or even damaged; a situation that is ethically problematic. Notice that the conclusion of the foregoing argument is not that new developments always *do pose risks*, but that they always pose *potential* risks, since the researchers cannot be familiar with all physically possible dangerous effects. In a word, insofar as a new technological development pertains to people or the environment, and is based on partial knowledge, the new development poses potential risks, and as such is ethically relevant.

Does the argument above also pertain to the innovative research into aerobic granular sludge sewage water treatment? In the first place, we do not need a lot of words to show the first premise also holds for innovative research in sewage water treatment; these new developments obviously are relevant for fellow human beings or the environment. It is clear from the outset that failure to meet primary or secondary effluent requirements could have serious effects on people and the ecosystem of the water treatment plant. The second premise, the partiality of the knowledge is a subtler issue. The first IWA workshop on aerobic granular sludge in Munich in September 2004 gave some important insights. We think it is fair to say that also in this field of biotechnological development, knowledge is incomplete. Some of the discussions indicate considerable differences of opinion. To substantiate our claim about the partial knowledge we mention some of those differences.

In the first place, participants did not agree on the mechanism that holds the granules together. Some researchers claim that lectins do the job, but this is denied by others, who claim that knowledge about this mechanism might even be redundant. Furthermore, it is unclear whether extra cellular Polymeric Substances (EPS) play an important role in the process of the granules formation. In connection with the uncertainty about the mechanisms, the question arises about the composition of the granules. The suggestion is made that granulation goes hand in hand with a change in the ecosystems of the competing bacteria, which favors hydrophobic bacteria that more readily aggregate in the human body. It is unclear whether the concentration of the various bacteria colonies of traditional activated sludge differs significantly from the concentrations of the granules grown in a sewage water reactor. Regarding the modeling of the granules formation, not all participants are convinced that granules can be modeled just as a biofilm, the way in which granules on carrier are modeled. Additionally, it seems unclear what role starvation plays in granulation. These theoretical differences of opinion are reflected in convictions of the engineers and scientists as to what research must be carried out to successfully build an aerobic granular sludge reactor. Some primarily want to concentrate on process technology whereas others favor microbiological research. Finally, some participants of the workshop doubt whether aerobic granular sludge reactor will ever be applied for municipal sewage water treatment, and believe the innovation to be only suitable for industrial purposes. The above account shows that the second premise of the argument also holds; i.e., knowledge regarding the aerobic GSBRs is partial. Consequently, the conclusion, drawn in general before, also holds for the research into aerobic GSBRs: it poses *potential* risks to people or their environment and is therefore, ethically relevant.

We want to emphasize that in the present research we focus only on ethical and methodological questions insofar as they relate to consequences of new technological developments for fellow citizens and the environment. In our parallel research, therefore, we will focus on the risks and hazards connected with bringing the Aerobic GSBR technology to maturity, which includes the required upscaling, and, eventually, the use of this technology.

One of the strategies to cope, is to develop and introduce a new technology in incremental steps. In the aerobic granular sludge research these steps consists of at least three different phases of the research, the laboratory scale, the pilot scale and the full plant scale research, where the latter again is divided into application at small subsidiary plants and full scale stand alone plants. We take the upscaling methodology to be a method to manage the relevant risks due to incomplete knowledge, and address therefore the following research question:

- What are the main risks inherent in the upscaling of aerobic granular sludge reactors?
- What are the ethical aspects of those risks?
- How are these risks influenced by (methodological) research decisions?
- How can research decisions on upscaling be improved with regard to the acceptability of the risks and hazard?

To be clear, the primary objective of the present research is not to find out how dangerous or hazardous the upscaling process is in this specific aerobic granular sludge research for sewage treatment. It is our goal to find out, how risks and uncertainties are handled, and how this is open to improvement. Secondly, and perhaps more importantly, we want to formulate a general scheme to analyze and procedurally improve risk management in multi-actor environments. This handling includes at least a serious attempt to chart the lack of knowledge by for instance, evaluating the discussions in the community of relevant researchers and a thorough investigation into the known risks and their management.

A research project as the one on GSB reactors is managed and financed in such a way that various people play different roles. In such complex situations it is common to distinguish different *actors* defined by different interests, knowledge and agendas. It is not one scientist in isolation who takes the relevant research decisions; the decisions are taken in a meeting of a board of various representatives. Thus, it seems to be inadequate to assess the research decisions from the point of view of an "overarching objective scientific rationality" only. To what extent research decisions may be called justifiable depends also on the information available to the relevant actors, their continuing relationships, and the need to shield decisions within the research project to some extent from "external" interests. In the sense that all of these matters may be legitimate considerations of actors, we need to take such considerations into account when assessing research decisions.

We have found out that different individuals among those knowledgeable about sewage water treatment did not agree about the extent to which certain risks are realistic, in the sense of "a factor to be reckoned with". Since we wanted to get an overview of the realistic risks related to the research GSB reactors, we asked a number of specialists in the field who represented all relevant actors. The goal of our paper is to report our findings and some observations about the risk perceptions of the actors involved.

In the next section we will explain our use of the Group Decision Room (GDR) to come to a list of risks considered to be relevant by the participants; after that we will tabulate the shortlist of risks brought up by the participants. In the fourth section we report on how the participants have scored the risks on two independent, five point scales pertaining to the probability and the impact of an event or process related to the introduction of the new technology. We focus, in the fifth section, on the question in which phase of the research the participants want the various risks of the shortlist to be addressed. After that, we discuss the problem of many hands, which, according to the data from the GDR, is far from being unrealistic, at least regarding some of the risks involved. We end with the conclusions and discussion.

Method: Group decision room (GDR)

To get an impression of what risks are important in the upscaling and introduction of aerobic GSBRs, we carried out a session in the GDR at the School of Technology, Policy and Management at Delft University of Technology. A GDR is an electronic environment for brainstorming and voting on issues. The GDR is sometimes used to help solve intractable policy issues by confronting the relevant actors with their different perceptions and looking for a possible consensus among the actors. We used the GDR, however, in a different way. The GDR session did not aim at a consensus among the relevant actors. Rather we wanted to trace potentially interesting agreements and disagreements about the risks of aerobic granular sludge reactors, and about the ways to handle them. We invited representatives of the various actors – such as researchers, engineering firms, users and regulators – to the session. We ended up with a group of ten participants, viz.:

- Three researchers, who are involved in the upscaling project for aerobic granular sludge reactors for sewage treatment, of which two were working at the Kluyver Laboratory, and one at the engineering firm DHV.
- Two researchers, who are not involved in the project. One of them was working at Paques and the Agricultural University in Wageningen. The other person was working at the Delft University of Technology, School of Civil Engineering.
- Four people from different water boards, i.e. the bodies responsible for sewage treatment in the Netherlands and as such potential users of the technology.
- One person from STOWA, the Foundation for Applied Water Research in the Netherlands. The water boards founded STOWA to finance research on new treatment technologies. STOWA is one of the financers of the upscaling project.

To address our first research question (see introduction), the goal of the GDR session was to find out the participant's thoughts about the risks involved in the development of the aerobic GSBR. The three parts of the session on which we will be reporting, were dedicated to the following questions:

1. What are – according to the participants – the risks of the research on, and the introduction of aerobic granular sludge reactors?
2. How do the participants assess the magnitude of these risks in terms of probability and impact?
3. In what phase of the research, design or use of aerobic granular sludge reactors should these risks be addressed?

Before we present and discuss the outcomes with respect to these three issues in the following three sections, we need to make two important remarks about the status of the outcomes of the GDR session as we see them. First, the outcomes of the session certainly do not amount to a risk assessment of aerobic granular sludge reactors. The session gave only an impression of what a number of specialists considered important risks. Second, the GDR session should not be thought of as a *measurement* of the exact perceptions of the people present. The main reason for this is the lack of objective criteria defining the words and concepts that were used. The aim of session was more modest. We wanted to get an impression of how people thought about the three above-mentioned questions.

A final word on the sample of people we choose might in place. Ten people can hardly be considered to be representative for the general perception of potential risks invoked by aerobic GSBRs. Although we have invited representatives of all actors involved, this is of course true, but – again – we did not aim at finding out what people generally or even what all people involved in sewage water treatment in general consider important risks of aerobic granular sludge reactors – in fact we would expect that most people would not be able to answer this question. Rather, we wanted to trace what the specialists in this area considered relevant risks. In order not to be too much biased towards what the people directly involved in the project consider relevant risks, we also invited some people involved in sewage water treatment but from outside this particular project.

What risks are considered relevant?

In the context of this project, we use the following definition of risk:

> The risk pertained to a new development is the product of the probability and the effect of an incident, or process, due to that development with undesirable consequences for people or the environment given:
>
> - the goals of water treatment;
> - the boundary conditions for technical installations in general.

This definition emphasizes the ethical dimensions of the risks, by mentioning undesirable consequences for people or the environment; one might think here of long-term or short-term health effects, reduced well-being, and damage to ecosystems. Undesirability is further focused on, by referring to the goals of water treatment and technical installations in general. For example, goals of water treatment are to prevent water-borne diseases, and to operate at acceptable costs. So if it turns out that certain pathogens are not eliminated, or that the costs to remove them are unacceptably high, then we should think of such situations or occurrences as risks in the sense relevant here. Apart from the specific goals of water treatment plants, there are certain occurrences that we would want to exclude for any technical installations, such as explosions or certain risks to those operating the plant.

In the GDR, we first presented our definition of risks to the participants. We then asked the participants to list as many potential risks as they could think of. This exercise resulted in longlist of more than eighty risks, which was reduced by the participants themselves to a shortlist of seventeen risks. After the session, we have classified these seventeen risks in six main categories. These results can be found in Table 1. The six categories of risks in Table 1 are conceptually distinct. Each category concerns a different type of risk. As one can see in Table 1, the six categories are in turn grouped into two main categories of risks: 1. risks *to* the introduction of the technology, i.e. circumstances or events that make it potentially difficult or impossible to get the technology accepted or implemented and 2. risks *of* the technology, i.e. potential undesirable circumstances or events that might occur once the technology is used.

Although the six risk categories in Table 1 are conceptually distinct, they are causally linked. Some of the mentioned risks can be a cause of some of the other risks. Although a plurality of routes of causation is thinkable between the six risk categories, the main causal routes seem to be the following. Risks related to the sensitivity of the treatment process, *C*, or risks directly due to upscaling, *D*, and the other risks, *F*, may eventually result in risks related to not meeting the requirements or secondary emissions, category *E*. If the primary – that is the legally obligatory – requirements cannot be met, or can only be met at high costs, for example due to additional treatment steps, then this in turn might lead to risks *to* the introduction of the new technology such as economical risks or the acceptance of innovative technology more generally.

Table 1: Risks

Risks *to* the introduction of GSBRs

A. Success/failure of this technology and effect on acceptance of innovative technology more generally
1. Failure of a new technology may adversely affect innovation in general.
2. Stricter effluent requirements complicate the choice for innovative technology.
3. Perhaps it will be difficult to get a license.

B. Economical risks
4. Eventually, the costs of the installation turn out to be too high.
5. The home market perspectives may be limited due to innovation fatigue.
6. The return on investment are too low, when launched in foreign market

Risks *of* the introduction of GSBRs

C. Sensitivity of the treatment process
7. The installation cannot cope with influent fluctuations (volume, composition, temperature).
8. Controllability of the granules formation.
9. The process is instable due to lack of self-regulation.
10. Expensive back-up mechanisms are needed in case the process stalls due to lack of robustness.

D. Risks directly due to upscaling
11. Certain factors can only be examined full-scale, not on lab or on pilot scale.
12. The pilot's local circumstances fail to be representative.

E. Meeting the requirements or secondary emissions
13. The process fails to meet the primary effluent requirements.
14. There are too much secondary emissions, in effluent or otherwise.

F. Other risks
15. Higher reactor vessels may cause the formation of foam.
16. Difficulties with sludge processing.
17. The operators lack competence.

Scoring the risks

After the participants of the GDR-session had made a shortlist out of the longlist they produced at the start, they had to *score* the seventeen risks of the shortlist. Having defined the risks as the product of the probability and the impact of an incident or process, we asked the participants to assign to every risk a number both on a five points probability scale and on a five points impact scale.

We have to make two remarks about the scoring exercise. Firstly, we must realize that the five points scales used are only ordinal scales. This means that if according to a participant event *16* receives an impact score 2 and event *17* an impact score of 4, this does not imply that the impact of *17* is two times as severe as the occurrence of *16*. This only holds for ratio scales. In our five points scale only the order of the events in the probability and impact dimensions counts, since the distances between two consecutive points need not be the same[1]. Secondly, the participants' scoring is of course completely subjective, therefore the outcome of the scoring exercise need not reflect the true state of affairs.

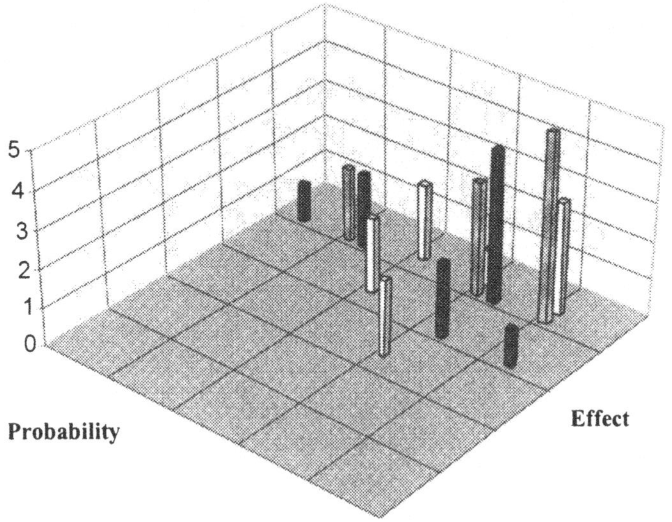

- The installation cannot cope with influent fluctuations (volume, composition, temperature)
- Controllability of the granules formation
- The process fails to meet the primary effluent requirements

Figure 1: Largest risks

An important consequence of this ordinal scoring of probability and impact is that the product of the scores is meaningless. We cannot, therefore, simply add the products of probability and impacts scores over all participants to calculate the largest risk. Since no event scores on probability and impact are the highest according to all participants, we can only make histograms to get an impression about what are perceived to be small and large risks. The two-dimensional histogram of Figure 1 shows the three events that received the highest sum of probability and impact values. These are: *the installation cannot cope with influent fluctuations (volume, composition, temperature),* the *controllability of the granules formation*, and *the process fails to meet the primary effluent requirements*. The first two pertain to the sensitivity of the treatment process, whereas the latter relates to the category of meeting the requirements. Without claiming that these three events are the largest risk pertaining to the aerobic granular sludge reactors, we do claim that these three events according to the participants pose real risks for the enterprise. In the same way, we may conclude from the data that the economical risk: *the home market perspectives may be limited due to innovation fatigue*, and a risk of the rest category, viz. *difficulties with sludge processing* are according to the participants the least hazardous. Figure 2 shows the two-dimensional histogram regarding the events that received the lowest sum of impact and probability.

[1] We even lack the means to fix the meaning of this clause, since the distance between two points is undefined.

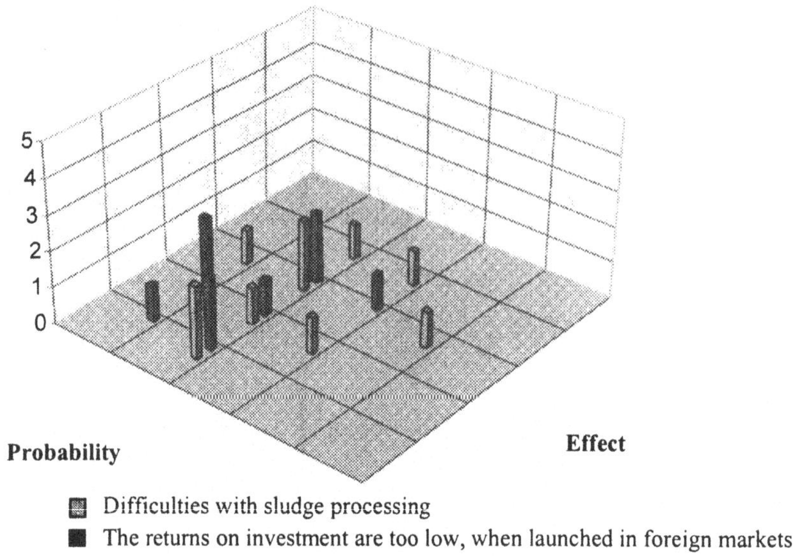

Figure 2: Smallest risks

Looking at the Figures 1 and 2, we see that the participants have significantly more worries about the question whether the installation can cope with the influent fluctuations than about a too low return on investments regarding the launching in foreign markets. Not only is the probability of the occurrence of the first event, according to most participants, higher than that of the second, also the impact or effect of the first is considered far worse than the effect of the second. The same holds for the comparisons with the other events depicted in the two figures. The GDR-exercise has, therefore, yielded significant differences between outcomes for different potential risks. Finally, another interesting observation reads that the dispersion of the values regarding events with high risks is less than the dispersion of the risk assignment for events considered to have a low risk. Consider, for example, the spread of values concerning the event that the installation cannot cope with the influent fluctuations. The estimates of that risk only conveys three, albeit not adjacent squares, whereas assessment of the risk that there will be difficulties encountered when processing the sludge covers eight squares. Whether this observation corroborates a more general tendency in risk perception is unknown to us.

The data on the estimated magnitude of the probability and the impact of the various risks show an interesting difference of opinion with respect to event *13*, the risk of not meeting the primary requirements, i.e. the requirements that are obligatory by law. All participants agree that the impact of not meeting this risk is large: three participants assess the impact of this risk as high and seven even as very high. There is, however, less agreement concerning the probability that this risk will indeed materialize. While seven people estimate this probability as either high or very high; three people estimate it as low or very low. What is even more striking is that two of the three researchers involved in the project and one of the other researchers – the one having experience with granular reactors – estimate the probability as low while all the people from water boards estimate the probability of this risk as either high or very high. The underlying pattern seems to be that the researchers are (far) more optimistic about the possibilities of meeting the primary requirements than the water boards, certainly if pre- and after treatment are provided. Time will learn whether the researchers are too optimistic or the water boards too pessimistic, but it is interesting to ask what might explain this difference of opinion and whether it is problematic for the further development of this technology.

One possible explanation may be a difference in knowledge: the researchers may have simply more knowledge of the research and the technology, on the basis of which they are more optimistic. This may be so, but one can doubt whether this is a completely adequate explanation. After all, some of the researchers are also pessimistic; moreover the people of the water boards who attended the GDR sessions are involved in the project and seem to have rather good knowledge of it. Another – perhaps complementary – explanation might be that researchers and water boards have different responsibilities. Researchers have a responsibility to develop new innovative technologies and doing so might require some optimism about the changes of eventually meeting the primary requirements even if this is not yet possible. Water boards, however, have a formal responsibility for water treatment and may have to face fines or other formal reprimands if they do not meet the legally required level of treatment. Given these stakes, pessimism about the possibilities of new technologies might be an understandable and apt attitude.

Finally, we will make a remark on risk perception. The assessments made by various people about the probability and the effect of the risks show that for five out of ten people, the assessment of the likeliness of the risk and that of the severity of the impact if the risk should occur strongly correlates ($r > 0.65$) (data not shown). Such a correlation is somewhat amazing. In general, the probability of a risk is not related to the severity of its impact. Consequently, we think that this correlation reveals more about people's perceptions of those risks than about objective features of these risks. Either people did not clearly distinguish between the concept of probability and that of impact, or they understood the difference but were not able to estimate them independently. In either case, people possibly first made an assessment about how serious they considered the event to be and from this, implicitly, derived an estimate of the probability and the effect of that risk, while keeping the estimate of the probability and the effect roughly the same.

Research on risk perception suggests that people, especially lay people, tend to include other concerns in their risk estimates that are, strictly speaking, not part of the magnitude of a risk defined as probability times effect (e.g. Slovic et al. 1990). In general, people tend to overestimate the magnitude of a risk if they consider that risk unacceptable. The acceptability of a risk, however, depends not only on the scale of that risk, defined by probability times effect, but also on factors such as: whether the risk is taken voluntarily, the expected benefits of the risky activity, the controllability of the risk, and the availability of alternatives. This means that people's estimate of a certain risk not only reflects the assessment of the magnitude of the risk, but also a judgment of the acceptability of that risk. The same probably holds for the participants of our GDR session, or at least a number of them. Remarkably, the correlation is positive in all but one case, whereas for five of the participants the correlation is high. For our purpose this is, however, not problematic. We will use the outcomes of the GDR session only to have a first idea about main risks that need to be considered during the development of aerobic granular sludge reactors. For this purpose, certainly not only the magnitude of the risk is relevant but also other factors that influence the acceptability of risks.

Where are the risks to be addressed?

The third question of the GDR-session was about when – i.e. at which of the stages in research, design and operation of the new technology – the seventeen risks had to be taken into consideration. For each of the seventeen risks of Table 1, the participants were asked to indicate one, and only one, phase or activity that they considered most important for addressing the risk. We defined ten activities or phases for this exercise:

- Lab research
- Upscaling to pilot plant
- Pilot plant research
- Upscaling to full-scale
- Mathematical modeling
- Full-scale Design
- Operation of full-scale plant
- New regulations with respect to full-scale plants
- New regulations on research
- Through contracts with suppliers

We mention some of the outcomes. First, the participants judged that most risks are to be addressed at the pilot-plant research. Apart from that, also full-scale design and the operation of the full-scale plant are considered to be important activities for dealing with the list of seventeen risks. Furthermore, according to the participants, the most important risks are to be addressed during the pilot research. Finally, regulations for research laboratories are thought to be irrelevant.

When we look at the three risks that were earlier defined as "largest" we see that at least half of the participants indicated that these risks should be addressed during the pilot research:

- *The installation cannot cope with influent fluctuations (volume, composition, temperature).* This risk was mentioned by eight of the ten participants as to be addressed during the pilot research.
- *Controllability of the granules formation.* This risk was mentioned by five of the ten participants as to be addressed during the pilot research. The opinions of the other participants were divided. Two participants delegated dealing with this risk to the lab research, one to the upscaling of the pilot plant to the full-scale plant, one to the mathematical modelling, and one to the operation of the full-scale plant.
- *The process fails to meet the primary effluent requirements.* Five participants conferred this risk to the pilot research. Again, the opinions of the other participants were divided. Two people allocated dealing with this risk to upscaling from lab scale to pilot scale, one to the upscaling of the pilot plant to the full-scale plant, and one as a risk to be dealt with through contracts with the supplier of the treatment plant.

On the basis of these findings we think that we are justified to conclude that in the perception of the participants the research at the pilot plant is crucial as far as addressing various risks of the enterprise is concerned. It would therefore seem important to make sure that the research setup allows for addressing these risks at the pilot stage, along with an assessment of whether the pilot stage is the most appropriate research phase, as some risks may only be addressable in later steps in the development.

Along with this conclusion we should add one caveat. One of the participants suggested afterwards that the concept "pilot plant" was not exactly defined beforehand, and it is unclear whether all participants had the same meaning in mind when deciding on the pilot plant questions. A pilot plant may be entirely inside a research laboratory with PhD-students being in charge or it might be outside the laboratory under the supervision of an engineering firm (see the contributions to the present volume of J.-H. Tay et al. and L.M.M. de Bruin et al.).

We think in our case the problem is small since we also included "Laboratory research" as a research phase. Participants who implicitly positioned the pilot plant research entirely in the laboratory would at least have mentioned the overlap between the two, in their conception. Moreover, the laboratory phase received overall very few votes for being the place where risks had to be addressed.

The problem of many hands

The answers given to the question "When are the risks to be addressed?" show also an interesting other result. Since the actors that were present at the GDR-session are also active in the research on, and design and operation of sewage water treatment plants, they implicitly assumed responsibility for themselves, and delegated it to others, by allocating activities to certain risks. If we look at the outcomes from this perspective, we see some interesting differences between the various risks. We especially look here at the risks: *difficulties with sludge processing* (event 16 in Table 1) and *secondary emissions* (event 14).

For the risks of difficulties with sludge processing, most actors allocate the responsibility for dealing with this risk to the phase in which they themselves are active. Four out of the five people from water boards – STOWA was included in the water boards here - indicate that the risks should be dealt with during operation of full-scale plants, i.e. the activity for which water boards are primarily responsible. Four out of the five researchers indicate that the risk should be dealt with during the pilot plant research, i.e. the phase for which they are responsible. For the risk of difficulties with sludge processing, we may say that almost all of the actors accept responsibility. This readiness to accept this specific responsibility is probably related to relatively limited perceived probability and impact of this risk.

The risk of secondary emissions shows a quite different outcome, however. Four out of the five people from water boards delegate this risk to the research phase, i.e. the phase in which they are not active and for which they are not primarily responsible. The picture for the researchers is less straightforward. Two researchers allocate this risk to the research phase, for which they themselves are responsible. Three researchers, however, allocate the risk to a phase for which they are not responsible. One researcher sees new regulations as a possible solution for this risk, which indeed seems relevant since the law does not regulate the secondary emissions – this was how the term secondary was defined in formulating the risk. This distribution might be due to the idea that it is not the task of those involved in developing this technology to worry about emission requirements not laid down by law.

This phenomenon of everybody allocating responsibility to others is akin to a problem that is known in the literature on moral responsibility as the "problem of many hands" (Thompson 1980; Bovens 1998). The problem of many hands usually refers to a situation in which something undesirable has occurred and in which it is very difficult, if not impossible, to pinpoint the responsibility for the undesirable event. This difficulty may be due to, initially, unclear allocation of the responsibility, or to the fact that many individuals contributed to the undesirable event. In the latter case, it might be quite difficult to find out who contributed to what extent to the undesirable event. Moreover, it may be very difficult to blame people for their contribution, for example because they could not have foreseen that their actions, or the absence of them, would contribute to the undesirable event. The situation we encounter here is somewhat different, since nothing undesirable has yet occurred. Nevertheless, the fact that all actors tend to delegate the responsibility for dealing with the risks of secondary effects to other actors may lead to the materialization of an undesirable risk and, subsequently, to the problem of many hands because it is unclear whose responsibility it is to address the risk and also, therefore, who is to be blamed once it has occurred.

Conclusions

We divided the risks mentioned by the participants of our GDR-session, into risks *to* and *of* the introduction of GSBRs. The first covered economic risks, and risks of acceptability; the second included the sensitivity of the process, failure to meet effluent or emission requirements, upscaling and other risks. In the opinion of the participants the following events pose real risks:

1. *The installation cannot cope with influent fluctuations (volume, composition, temperature)*;
2. *Controllability of the granules formation*;
3. *The process fails to meet the primary effluent requirements*.

According to the participants the following two events are the least hazardous:

1. *The home market perspectives may be limited due to innovation fatigue*;
2. *Difficulties with sludge processing*.

According to most respondents the main risks are to be dealt with mainly during the pilot research, the phase which also overall took the most votes as far as addressing the identified risks is concerned. The design and the operation of the full-scale plants take a good overall second and third position. Whereas the participants took some risks to be their own responsibility, they regarded other risks not to be of their concern. The risk of not meeting the secondary emissions serves as an example. The problem of many hands occurs if an event takes place for which nobody feels or is appointed to be responsible. In the GSBR-case this would happen, for instance, if after years of operation it came out that the pathogenicity of the effluent turned out to be unacceptably high.

Acknowledgement

We thank the *The Technology Foundation STW,* and the *Netherlands Organisation for Scientific Research (NWO)* for the grant that enables us to carry out this research. We would like to thank all participants in the GDR session. We also want to thank Ir. C. Uijterlinde of STOWA, Ir. B. de Bruin and Ir. H. van den Roest of DHV for their cooperativeness and the members of the supervisory committee of the project, who welcomed us to their meetings; and last but not least we would like to thank Ir. M. de Kreuk and Prof. M. van Loosdrecht for the time and determination to give us some insights into the mechanisms of the Laboratory research and the Pilot Plant at the Ede municipal water treatment plant.

References

Bovens, M. (1998). The Quest for Responsibility. *Accountability and Citizenship in Complex Organisations*. Cambridge, Cambridge University Press.

Bruin, L.M.M de, H.F. van der Roest, M.K. de Kreuk, M.C.M. van Loosdrecht (2004). Prominsing results pilot research aerobic granular sludge technology at WWTP Ede. This volume.

Tay, J.H. et al. (2004). A Comparative Study of Aerobic Granulation in Pilot- and Laboratory-Scale SBRs. This volume.

Slovic, P., B. Fischhoff & S. Lichtenstein (1990). Rating the risks. In: T.S. Glickman & M. Gough (red.) *Readings in risk*. Washington: Resources for the Future, 61-74.

Thompson, D. F. (1980). "Moral Responsibility and Public Officials." *ASPR* **74**: 905-916.

Discussion Outcomes

M.K. de Kreuk[*], B.S. McSwain[**], S. Bathe[**], S.T.L. Tay[***], N. Schwarzenbeck[**] and P.A. Wilderer[**]

[*]Department of Biotechnology, Delft University of Technology, Julianalaan 67, 2628BC, Delft, The Netherlands (E-mail: m.dekreuk@tnw.tudelft.nl)
[**]Institute of Water Quality Control and Waste Management, Technical University of Munich, Am Coulombwall, 85748 Garching, Germany
[***]Environmental Engineering Research Centre, School of Civil and Environmental Engineering, Nanyang Technological University, 50 Nanyang Avenue, Singapore 639798

Introduction

The outcomes from the discussions that followed each session are summarised in this chapter. The discussion outcomes are collected into one chapter, since most subjects were discussed several times during the workshop and extra information gained throughout the sessions in combination with extra time for refreshing thoughts, led to the development in the total discussion. The discussion outcomes are separated to research subject and this chapter starts and closes with the outcomes of the final discussion.

The closing session of the aerobic granule workshop started with one of the most important topics of the workshop: what have we been talking about? In other words, how would we define an aerobic granule? In literature, biofilm particle, sludge floc and aerobic granule is often mixed up, so it is time to define a general description of a granule. After a long discussion it was decided that:

Granules making up aerobic granular activated sludge are to be understood as aggregates of microbial origin, which do not coagulate under reduced hydrodynamic shear, and which settle significantly faster than activated sludge flocs.

This means that SVI_{10} (SVI after 10 minutes of settling) in combination with SVI_{30} should be used for characterizing the settleability of granular activated sludge as was suggested by Schwarzenbeck et al. (2004). Both values should be reported in papers written about aerobic granules, since the difference between the SVI_{10} and SVI_{30} value gives an excellent indication about the granule formation. Sieving is considered a proper method to harvest granules from activated sludge tanks or from aerobic granule reactors.

Other characteristics of aerobic granules that were mentioned by the delegates were:

- Aggregate were the position of micro-organisms does not change quickly as in an activated sludge floc, where an organisms can be in the centre at one moment and on the outside the next moment. The structure of a granule is determined in a matrix of biomass and EPS.
- No carrier material is intentionally involved or added, the aggregate is formed without the dosage of such carrier material. This item is rephrased to "aggregate of microbial origin".

- No thickening after settling or rapid settling of the aggregates. This was summarized in the part "which settle significantly faster than activated sludge flocs" and the demand to report SVI_{10} and SVI_{30} when aerobic granules are described in literature.
- In addition, the minimum size of granules was discussed, in order to be able to define the ratio of granulation by the determination of particle size distributions by image analysis or sieving. The minimum diameter that is needed was not agreed on, however granules are considered harvestable by sieving, which determines the strength of the required biomass matrix as well. Sieving can also be a suitable method for determining the granule content of the reactor, expressed as percentage of the total biomass. When sieving is used for determining size distributions, mesh sizes are needed to be reported.

When an aggregate fulfils all characteristics as described above, it can be called aerobic granular sludge. With this definition, the delegates have tried to set a standard for future research on aerobic granules. This simplifies the interpretation of experimental results and clarifies when to speak about aerobic granule formation and when about activated sludge or biofilms.

Aerobic Granule Formation

Since the stability of granules could be compromised if suspended flocs were allowed to proliferate, it was important to understand the factors that drive the dominance of granules over flocs and vice versa, and develop operational strategies to enhance the stability of aerobic granules. Some delegates were able to maintain stable acetate granules even after two years of operation and did not observe any instability problems. Other delegates encountered episodes of granule disintegration, for instance, if no sludge wasting was carried out. Differences in experience led to many discussions about the important factors for granule formation and granule stability. Existing literature on aerobic granule sludge typically focuses on a few parameters that influence granule formation:

- Feast-famine as provided by the SBR operation and other substrate dependencies in combination wit growth rate;
- High shear force;
- Short settling times;
- EPS formation;
- Inclusion of divalent cations.

Most of these parameters were discussed during the workshop and are separately summarized below.

Substrate type and concentration in combination with growth rate

During the workshop, many explanations for granule formation were given. In the pilot plant research selection for slow growing organisms (mainly phosphate-accumulating organisms) by dosing the influent anaerobically, is the main selection criterion. Since it is possible to select for the slow growers, other parameters as shear and settling velocity, which are important in combination with high growth rates become less relevant. Studies at the TU Munich - Germany, with different types of wastewater, showed that the morphology of the granules highly depends on the composition of the wastewater (and thus complex substrate) they were grown on. As an example, Schwarzenbeck presented results of an investigation on the treatment of dairy effluents in an aerobic granular sludge reactor. During the experiments structural instabilities of the granules occurred due to filamentous outgrowth. The growth of filamentous organisms could be related to the fact that in dairy wastewaters metabolically available substrates are slowly released at low concentrations due to hydrolysis.

It was generally agreed that carbohydrates and low dissolved oxygen levels led to flocculation, due to the favoured growth of filamentous organisms. Moreover, other independent studies had previously shown that carbohydrate-fed granules contained substantial amounts of filamentous microorganisms while acetate-fed granules were usually dominated by floc-forming bacteria and were generally more compact. Also Liu found in his experiments that the use of sucrose as sole carbon source produced granules that contained filamentous microorganisms while use of a dual substrate of sucrose and acetate resulted in granules without filamentous microorganisms.

Van Loosdrecht presented that experimental observations indicate that the higher the potential growth rate on certain substrates the more difficult it is to get granules. For example, it is easier to get compact structures on methanol than on acetate, because of the different growth rate of the organisms on these two substrates. The only observed exception was growth on glucose. Normally, microorganisms have fast growth rates on glucose, but in biofilm and granule systems, populations are found that have low growth rates on glucose and therefore form dense and smooth structures. As example, a wastewater was discussed, which contains formate and acetate. It is expected that under the right circumstances there is no problem for granule formation on this influent. However, since one substrate is a C1 substrate and the other an "easy" C2 substrate Van Loosdrecht postulated that one substrate could be used for growth and one for energy supply, complicating the evaluation of such system. This might need extra attention for the right circumstances.

Besides the composition of the influent, the method of dosing it to the reactor effects granule formation and stability as well. As an example, sequencing Batch Reactor (SBR) operation versus continuously fed reactor operation was discussed. A feast-famine regime in a SBR is considered to favour granule formation. The definition of "feast-famine" and starvation in a reactor used in aerobic granule research is different from the definition microbiologists often use. Therefore, it was clarified what the delegates mean with a feast-famine regime. It was noted that "famine" and "starvation" are generally used to describe the absence of external substrate in a reactor. However, these terms are misleading since bacteria can be growing on stored substrate products (PHB, glycogen, etc...), as generally occurs during the majority of the SBR react phase, after the initial substrate uptake. From a microbiological perspective, this is a "growing" phase rather than a "starvation" phase. Starvation only occurs when all stored products are metabolised.

Despite the misleading terms, feast-famine is used to describe the variation of substrate concentration in the reactor volume.

Researchers agreed that the variations in feast-famine provided by the SBR favourably influenced granule formation. A previous study by McSwain et al. was cited as showing that decreased feast-famine in the SBR led to decreased granule formation and stability. However, Tsuneda et al. pointed out that (nitrifying) granules were formed in his reactors under continuously fed operation. Alternatively, an anaerobic feeding phase had a profound effect on granule stability. Several delegates advocated feeding at the reactor bottom with anaerobic fill as opposed to feeding from the reactor top. The first approach ensures anaerobic conditions, which can enhance granule stability by selecting for slow-growing glycogen accumulating organisms (GAO) or phosphate accumulating organisms (PAO).

In the discussion of type of substrate and method of dosage, growth rate of the organisms played a central role. Therefore, a summarizing conclusion might be that not the substrate as such is the determining factor for granule formation, but the rate at which the organisms can grow on it is.

Role of shear

Shear is a parameter that forms the base of many discussions. Granule formation on acetate in a bubble column was not possible in the experimental set-up as used by the researchers at the TU Delft, The Netherlands. These reactors have a height over diameter (H/D) ratio of 15. Mosquera-Corral managed to form granules on acetate in a bubble column, but in the experiments in which pulse feeding was applied, H/D ratio's were about 50% smaller. Therefore, the stratification in these reactors is smaller and mixing better. With high H/D and thus high stratification, the top granules face low shear stress (depending of course on aeration flow rate), and will grow out more easily with filamentous and/or fingertype structures (no granule formation) as is often observed in continuous fluidised bed systems as well. In this case shear is an important parameter for the control of the outgrowth of the fingertype structures.

In experiments with anaerobic feeding and thus with the presence of PAO or GAO, stable granules were found in a bubble column and high H/D ratio's, as well by Schwarzenbeck as by de Kreuk.

Liu Y.H. discussed how superficial upflow air velocity and type of carbon source affected the stability of aerobic granules in a Sequencing Batch Airlift Reactor (SBAR). At a low superficial velocity, an outbreak of filamentous microorganisms was observed that gave rise to a poorly settling sludge and eventual biomass washout. When a higher superficial velocity was used, granules with significantly improved settling characteristics were obtained.

In most discussions about shear, the difficulty of measuring shear plays a role as well as the oxygen concentration, which is often varied at increased superficial upflow air velocities. To be able to draw conclusions from experiments, these factors need to be controlled and standardized as proposed later in this chapter.

Short settling times

Strategies to limit the amount of flocs in an aerobic granule system include use of short settling and discharging times. It was recognized that the selection pressure imposed by short settling times should be more important in fully aerobic systems, but in anaerobic-aerobic systems with phosphate accumulating organisms (PAOs), the settling criteria seemed less important because of the inherent tendency of PAOs to aggregate.

Role of EPS formation

Aerobic granule literature reports conflicting data regarding extracellular polymeric substances (EPS) and aerobic granule formation. Researchers reported various results regarding increase in EPS amounts with granulation, differences in loosely bound and tightly bound EPS, and insoluble versus soluble polysaccharide gradients within the granule structure.

EPS could be the glue between the organisms of an aggregate and EPS may have important functions with respect to cell metabolism. Examples are that exo-enzymes are often an integral part of the EPS enabling cells to get access to substrate of larger molecule size and that EPS could be important to protect cells against toxic substances because of its absorptive capacity. Van Loosdrecht agrees on the importance of EPS and that more research is needed before conclusions about the role of EPS can be drawn. However, he doubts that

EPS is needed only as glue, in that case nature would make a much more efficient material and glue the cells together with much less polymer. It is more likely that the EPS is produced as an integral part of the microorganisms and can therefore initially be included as part of the normal growth process.

During the different discussions about EPS, no consensus among the delegates about this subject was reached. Few of the major questions that were left were: how can shear force influence bacteria to form EPS within a granule, when shear force primarily impacts the boundary layer and there are steep substrate gradients within the granule and what is the exact role of the EPS and why is it produced in the granule?

Influence of other additions

The final parameter influencing the granule formation regarded the addition of divalent cations such as iron and calcium to aid granule formation. The work by Tsuneda et al. suggested that high concentrations of cations might increase the rate of granule formation and might influence the stability of the reactor system. Researchers discussed what concentrations of cations were required, and if this approach could aid full-scale formation and start-up.

All different parameters that were discussed seemed to play a role in the granule stability. There was no concluding agreement on which of this parameters is the crucial one and most probably it is a combination of all. In the studies on aerobic granule stability parallels to biofilms and bulking sludge can and should be drawn in order to increase the understanding of this system. Combining all experimental and modelling knowledge about these different, but related morphologies might lead to a better understanding about which are the crucial parameters for aerobic granulation and which are minor side effects. Also communication between research groups as during this workshop might help to reveal the uncertainties of the granule formation mechanism.

Biodiversity and population control

Aerobic granules can be considered as mini-ecosystems, consisting of a mixed microbial population. By determining the right conditions, one can try to manipulate the dominant organisms. A mixed population system often raises questions about which organisms are dominant, which are the most important for the conversions and structure and if we can add the organisms that we would like to be present. The discussion about biodiversity focussed on the possibility of adding or the possibility of existence of certain organisms. Moreover, the analytical techniques to investigate biodiversity were discussed.

In the granule formation discussions, it was stated that the selection for phosphate or glycogen accumulating organisms is advantageous for the stability of the granules. A positive side effect of the selection of PAO is the biological removal of phosphate. However, competition between phosphate and glycogen accumulating organisms in the anaerobic/aerobic operated SBR was questioned. This competition is known to be sensitive towards temperature and SRT. According to Zhu this competition is a difficult topic, since not much is known yet of the sensitive balance between those two species. Zhu focused on good granulation and reported that the PAO/GAO were easily maintained in the system.

De Kreuk operated a reactor with dominantly PAO, however FISH results showed that GAO co-existed in these granules. PAO stayed dominant as long as P was available in the influent and SRT was not too long (<30 days, temperature 20°C). When influent with limiting amounts of phosphate was dosed, granules dominated by GAO could be grown at

20°C. Understanding this competition is important for large-scale applications in different climates. Therefore, additional research is needed.

Another question towards biodiversity was raised; would it be possible to grow Anammox inside nitrifying granules. Researchers agreed on the theoretical possibility of Anammox in the anoxic core, since the SRT of the granules is long and especially the core of the granules can become very old, effectively the CANON process is based on such granular sludge. However, when nitrifiers in the outer layer are to active, ammonia will not penetrate to the interior and Anammox cannot grow. If ammonia can penetrate the core of the granules due to high concentrations growth of Anammox is possible. Mosquera-Corral and de Kreuk both checked their systems on the existence of Anammox and both researchers did not see any, even after inoculation with Anammox by de Kreuk.

The overall changes in biodiversity can be investigated by microbiological techniques as DGGE. Tay reported on the feasibility of using acetate-fed granules as a microbial inoculum for phenol biodegradation in aerobic granular sludge reactors. Steady state conditions in terms of MLSS-content could be reached within two weeks of operation; steady state in terms of phenol degradation could be reached within three days of operation. From the DGGE band patterns he concluded, that sudden phenol addition had no influence on the microbial diversity in the reactor and at the same time on the functional diversity. Nevertheless, a clear shift of the band-pattern towards different organisms compared to the inoculum could be observed from the DGGE-gel.

McSwain observed that DGGE profiles from aerobic granules contained few dominant bands, which did not show much variation over time. In contrast, profiles produced from a predominantly flocculent sludge contained a higher band number and showed temporal variations. This leads to the following question: do microbial populations of aerobic granules have a higher stability than those of flocs? Connected to this question are the following problems: i) Full-scale reactors usually have a more diverse, but also more stable microbial population than laboratory-scale systems. Will the above raised question then also be valid for full-scale systems or is it a laboratory artefact? ii) Is stability of the microbial composition necessarily related to granule stability and stability of the system performance? These questions form nice research questions for the future and were not answered as such during the discussion sessions, since specific knowledge about this was lacking. These questions resulted in a discussion about the use of microbiological techniques in aerobic granule research as is described in the next section.

Analytical Techniques

Molecular Microbiology techniques

In general, the question was raised if environmental engineering sciences are able to profit from molecular microbiological investigations in terms of system design and overcoming performance problems of existing reactors.

There was a general agreement that the costs and time requirements of microbiological investigations would exceed a feasible limit for the handling of problems in existing systems. Most of the present researchers agreed that parallel microbiological research should be continued to improve scientific knowledge, which in the long-term might help to design treatment systems which are better adapted to the physiological requirements of the microbial (sub)populations involved in performing the task of interest. The presented microbiological techniques can mainly be used for understanding operational problems of an engineering system ex-post.

It was also agreed that 16srNA based profiling methods such as DGGE are not an adequate tool to determine functional diversity in a system and to give detailed information on the structure of the microbial community without performing further functional (e.g. enzyme activity) and phylogenetic investigations.

FISH

From the FISH results presented by Zhu, the question was raised how to use FISH techniques in granule systems. Ivanov mentioned that the use of positive and negative controls is very important, since a hydrophobic probe can stain any hydrophobic part of the granule and this might give unspecific signals. According to Bathe, similar problems with regards to activated sludge have been overcome in a variety of studies, e.g. in activated sludge research. The washing step is very important and when this is done with care, the unspecific binding can be greatly avoided.

The problem of micro-organism density and penetration of the probes and of the fluorescent light is another problem that has to be overcome. Generally, it was agreed on that with CSLM it is possible to distinguish between organisms and debris. Also, the use of FISH for primary examination can be very useful, when a general idea is to be formed about the groups of organisms that are present.

Measuring Shear

The discussion about the influence of shear on granule formation also led to the question if the average shear should be calculated and if that is useful. Quantification of shear was found to be relevant for the comparison of results in granular sludge studies. However, the determination of effective shear stress is difficult, since that depends on many factors, as morphology and density of the granules determine the impact force the granules face. Van Loosdrecht remarks the influence of particle-particle interactions has been investigated (in Gjaltema *et al.*, 1997). The smoothness of biofilms in continuous airlift biofilm reactors is balanced by carrier concentration and the collisions they make with the biofilms. Gjaltema et al. (1997) found that the radius of the particles is more important for detachment than the density of the biofilm (power 5). An increased biofilm thickness led to higher impact and detachment, so particles maintain a balanced size. An interesting observation in this research was that instead of spherical particles, egg shape particles were formed at increased shear. That means that there are zones with higher detachment even on 'spherical' particles. It is unknown what that means for the distribution of microorganisms (nitrifiers and heterotrophs). This could be an interesting topic for further research.

Measuring EPS

Since many studies of the role of EPS are published and the role of EPS is still unclear so more research on this topic needs to be done, it was generally agreed that researchers should clearly describe extraction and characterization methods for EPS, since much data cannot be compared due to methodological differences.

Measuring and characterizing the EPS in granular sludge is considered important by the researchers. The most appropriate methods for characterizing EPS in granular sludge is however not standardized yet.

Conversion processes

Because of the unique structure of aerobic granular sludge, with the presence of COD in the total granule in combination with anaerobic and aerobic zones, simultaneous removal of most components of wastewater can take place. This structure has been studied over the past years and the structure related to conversion processes in the SBR's was often discussed during the workshop.

Ivanov presented data about the internal structure of granules, showing a layered structure of granules, with aerobic heterotrophs and obligate aerobic autotrophs in the outer layers, followed by a layer with reduced oxygen tension in which facultative anaerobes are situated. Finally, the existence of obligate anaerobes proved the existence of completely anaerobic zones within granules. Above a certain diameter, dead biomass is going to accumulate within granule centres.

The access of substrate to the inner layers of a granule is mass transfer limited. To which extent the existence of pores and channels enhance this mass transfer is still unknown. The question was raised if these organisms use only products of cell lysis for growth or if they contribute to the overall substrate removal (e. g. of anaerobically degradable substances) of a reactor system as well, considering mass transfer limitation of substrate. The same question should be addressed considering the possible occurrence of simultaneous nitrification and denitrification in outer and inner layers of granules, respectively.

Related to the mass transfer questions (and to the question of stability of microbial composition) is the question of optimal granule diameter; which would be a compromise between a high settling rate and a maximal proportion of metabolically active biomass within the granule. Should the granule diameter be constant for optimal operation or is a natural fluctuation of granule diameters within a certain range to be expected and allowed? Additionally, which would be the best way of "diameter control" through the operational parameters of the reactor system? Most probably, one can affect the granule diameter with the same parameters as are influencing the granule formation. Nevertheless, how this can be done in a controlled way is not known yet.

Mosquera-Corral discussed an aerobic granular sludge SBR that was operated at different COD/N ratios. Removal of ammonia was shown to occur via assimilation and simultaneous nitrification/denitrification. The ratio of those two processes depended on the COD/N ratio in the influent. During the first 5 experimental periods the COD concentration was maintained, while the ammonia concentration was changed, while during the last 3 periods the COD concentration was decreased to 0, while the ammonia concentration was increased. During the last period (NH_4^+-N concentration in the influent 100 mg/l) granules broke, with as possible causes the disappearance of the heterotrophs from the granule structure or the higher free ammonia concentration or pH fluctuations. However, in time stable nitrifying granules were obtained under fully autotrophic conditions.

The calculation of conversion from mass balances used in these kinds of studies was discussed. Researchers agreed on the difficulty of balancing because of the accuracy of the tests used and the accuracy of the influent dosage, however that is not different from other wastewater systems. Interpretation of the results and presentation of the concentrations in SBR's is a bit more complicated but is needed. The effluent that is not removed, since it is below the effluent extraction point, dilutes the influent. Furthermore, the nitrite or nitrate present in this water is first denitrified directly after feeding with the dosed COD. The nitrogen removal efficiency is therefore related with the concentrations in the dosed influent, the nitrogen that remains in the reactor after effluent withdrawal and the amount removed in the aerated phase. Balancing the nutrients in this kind of systems is possible when also the off-gas from the reactor is analysed. Biomass production can be measured during long term of steady state operation.

Zhu discussed the improved phosphate removal capacity of aerobic granular sludge in comparison to flocculated sludge. Bio-P release was observed in the first part of the anaerobic period. Fast phosphate uptake took place during the initial stage of the aerobic period. The results demonstrated that biological phosphate removal with aerobic granular sludge was possible and might have contributed to the formation of the granules and its structure.

The start-up of bio-P systems was discussed. The start-up of the system of Zhu took about two months and after 5 months a stable steady state system was reached that was operated for more than 2 years. Dulekgurgen operated a bio-P granular sludge reactor too, in which granules were present after 1 month of operation on an acetate/butyrate mixture. However, bio-P activity disappeared after some time and inoculation with new EBPR sludge did not give expected results, since all sludge washed out quickly. Zhu suggests the dosage of new sludge as well when the granules disintegrate and while the reactor is operated as during a new start-up. De Kreuk operated bio-P granules as well. These granules also came in existence in one month and were fully grown in two months. This system is operated for over 2 years now and is still stable. The settling time was adjusted during start-up to keep the EBPR sludge in the reactor and was shortened when the first granules appeared.

Presented results and discussion in this session reported a great possibility for simultaneous removal of different components from the wastewater in one reactor, because of the structure of aerobic granules. However, stability of granular biomass and granule diameter and its influence on the conversion processes has to be investigated more, since this might be an important factor for large scale applications and knowledge about influencing granule diameter is lacking.

Modelling

Lübken used the activated sludge model No. 3 (ASM3) in order to model an aerobic granular sludge sequencing batch reactor, treating malting wastewaters. He concluded that, although ASM3 was originally designed for modelling suspended growth systems, it can be used to describe biofilm moving bed systems (e.g. aerobic granular sludge reactors) when adjusting the model parameters adequately. As in the investigated system a high level of protozoa growth was present, for instance the heterotrophic decay rate was increased.

This led to the following general questions:

- Is it permissible to use a model, that was developed for suspended growth for systems with suspended biofilms, where for instance also diffusion processes have to be considered?
- Is it necessary to develop the most possible complicated model to get good modelling results?
- How can the role of protozoa be included in the existing models?

It was discussed in detail, whether adjusting model parameters beyond those values that have been found in the past can be used to transfer a model for suspended growth to a system with attached growth. It was agreed that developing an extremely detailed and complicated model is not essentially necessary, if simple models give good correlation between the experimental results and the calculated data. Further it was agreed, that the role of protozoa is highly neglected in the existing models, although they contribute to the decay rate (grazing predators) and the removal of inert material (sessile filter feeders). Schwarzenbeck reported on his experiments with malting wastewater that contains a high portion of particulate matter. In that case a very dense population of sessile protozoa was found that could be identified responsible for uptake of particulate matter from the wastewater using tracer particles. Further research is needed, although there are some but very few studies in the past that indicate an important role of protozoa.

Pilot plant operation and full scale design

Investigation of the new aerobic granular sludge technology is not only done out of a scientific interest, but moreover, it can add a compact, simple and robust technology to the wastewater treatment systems. Experiments are performed so far under well-controlled laboratory circumstances and the full-scale technology is limited to the design of such a plant on paper. Most delegates had their ideas about a full-scale aerobic granular treatment plant and during the workshop this aspect was brought up for discussion once in a while.

The discussions focused on the presented first pilot plant research in Ede, the Netherlands and on up-scaling of the technology in general. Topics discussed included the difference between existing systems and granular sludge technology, difference between lab-scale and pilot, economics, energy and expected disadvantages of this technology.

First pilot plant research on a sewage plant

De Bruin presented the first results of a pilot plant study on sewage at the treatment plant in Ede, The Netherlands. This pilot reactor has a height of 6 meters; a diameter of 0.6 meters and the hydraulic capacity is approximately 5 m^3/h, depending on the applied load. The operation of the pilot plant during the time prior to the workshop focused on getting granules and not on nutrient removal, behaviour of suspended solids or SRT control. These factors will be investigated as soon as granule formation is completed. For example, SRT will be controlled after start-up by actively removing granular sludge. The granules will be studied for handling and the sludge production will be measured. In this study, also the aspects and differences of post-treatment and pre-treatment will be investigated. A parallel study with wastewater will be done in industry and soon, a full-scale plant will be started up in the Netherlands in food industry.

Start-up of the pilot plant took longer than in the laboratory scale experiments. The experiment in a laboratory study with sewage as influent needed a start-up period of 6 weeks. In laboratory research, the circumstances can be well-controlled and optimised, as for example pH, influent and temperature. The influent temperature in the pilot plant had to be controlled in winter, since the total surface area/volume ratio of the pilot reactor is much higher than it would be in practice. This made starting-up in winter with Dutch winter temperatures difficult. In laboratory research it was shown that starting up at 8°C was impossible, whether going from 20°C to 8°C did not affect granule morphology. At the time of the workshop, this experiment at low temperature in the laboratory lasted already 2 sludge ages (≈4 months). Since winter doesn't last longer than this in the Netherlands, no problems are expected for stable operated granular sludge systems.

The question was raised if stratification in an SBR-bubble column of 6 meters height could play a role in granule development. To avoid this an airlift reactor should be used. However, the effluent extraction takes place at half the reactor height. Holes in the riser could prevent the effluent flowing from the riser to the downcomer via the settled bed, but these could affect the flow pattern during aeration as well, with unknown consequences for granule development. Using an airlift as high as the effluent extraction point is another solution, but than half of the reactor is well mixed and stratification will still occur. When granule formation is complete, stratification (for size and density) in the reactor will be investigated.

Full scale design

Different factor that are important for granule formation, will also influence the full-scale system, both on economics as well as on construction. In some of the discussions on granule formation and conversion processes, the influences of the discussed parameters on a full-scale plant were debated.

Shear forces were considered to be important for granule formation and obtaining smooth and dense granules. However, high shear force is typically supplied by high aeration rates. Workshop delegates acknowledged that the supply of high shear and high dissolved oxygen concentrations are likely to result in major costs for full-scale aerobic granular SBR operation. The significance of shear force to aerobic granule formation, for both aerobic and anaerobic feeding conditions was discussed. In this respect, it was suggested that the use of anaerobic feeding period diminishes the need for high dissolved oxygen concentrations and high shear force and therefore can reduce investment and operational costs.

Another aspect influencing a full-scale reactor is the way of feeding it. Pulse feeding can give problems like high concentrations of free ammonia and pH control difficulties in the studies. This method does give insight in the process of polymer storage and has practical advantages at laboratory scale, yet, dump-fill is not logical at large scale applications and also sludge stability can be solved by a longer anaerobic feeding period as was shown in the presentation of Zhu.

The discussion on the applicability of aerobic granular sludge technology to the treatment of slowly degradable substrates was less unison. If the wastewater is provided at a COD-loading rate that allows only very slow substrate uptake, the SBR-system moves more towards a continuous fed system and sludge granulation becomes impossible, filamentous organisms grow. On the other hand other researchers believe that this does not contradict an application to slowly degradable wastewaters as an operational strategy can be chosen that allows the selection for slowly growing organisms, which again lead to the development of smooth granules.

Another critical aspect recognized for large scale applications is that the commonly applied operational strategy of washing out slow settling biomass together with the effluent in order to select for fast settling granules, creates an effluent that can be impossibly discharged into receiving waters without further post-treatment. It was agreed that a separate discharge is essential in order to obtain dischargeable effluents. Various strategies were discussed:

- Allow complete settling of the sludge bed and discharge excess sludge at the top layer of the sludge bed, assuming that the sludge bed shows stratification of granules at the bottom and flocs on top.
- Diverge the first couple of percent of the effluent into a settling tank, as this contains the biggest part of MLSS in the effluent. The rest can then be discharged into receiving waters.
- Discharge during aeration and treatment in a separate settling tank

It was commonly decided amongst the researchers, that complete settling in the SBR and discharge of the excess sludge from the top of the sludge bed is the most practicable solution although it would require more advanced control of the sludge discharge point.

From the debates about the aspects influencing the full-scale design can be concluded that operating the system via an anaerobic feeding period, followed by an aeration period, can solve many disadvantages. As is also shown in laboratory experiments, selection for slow growing organisms leads to stable granules at different process conditions. The effluent characteristics and its suspended solids concentration should be investigated at pilot scale, since the influent can strongly influence the solids in the effluent.

Comparison with other systems

Feasibility of the aerobic granular sludge technology depends on the competing systems in aerobic wastewater treatment. Compared to other systems, it should be cheaper, more robust, simpler, or have other characteristic advantages. The discussions about the differences with existing systems are summarized below.

Aerobic granular sludge reactor (GSBR) technology differs from existing systems as the biofilm airlift reactor (e.g. the Circox reactor). The GSBR is a simpler design, since aerobic granular sludge can be obtained in a simple tank wit a bubble aerator. The use of an airlift is not needed with the Bio-P granules. Also the operation is more stable when easy degradable substrates are present in the influent. In a biofilm airlift reactor, the basalt concentration determines shear and as discussed in earlier sessions, the biofilm morphology in these reactors. In buffer tanks and sewage systems easy degradable fatty acids are produced, which can quickly lead to unstable situations. Furthermore, biological phosphate removal is easier achievable in a SBR. The internal anoxic zone in the airlift reactor for the nitrification/denitrification mechanism and the three-phase settler of a Circox are more difficult to construct than the straightforward SBR. Finally, variations in hydraulic loads are easier to handle in a SBR system than in a continuously operated biofilm airlift reactor. In the latter variations of maximum around 20% to 30% can be dealt with, while the SBR can handle complex different and fluctuating hydraulic loads. The granular SBR system has of course the disadvantage of more complex planning due to the discontinuous nature of the process

The differences with conventional systems were discussed as well. Disadvantages of the GSBR system are the start-up of the reactors. Where activated sludge reactors are easily and quickly started-up, GSBR systems need a longer time or a start with adding of granules from another installation. Also the robustness and stability of aerobic granules in practice is an uncertain aspect. However, this will be found out at pilot scale.

Aeration per COD load is expected to be similar to conventional systems. Probably the aeration equipment to tank volume has to be bigger, since the COD load in GSBRs will also be higher. Furthermore, energy costs are expected to be lower. In a GSBR, no energy is needed for mixing devices in anaerobic or denitrifying tanks, pumping of return sludge or operation of sedimentation tanks. A feasibility study pointed out that as well the investment costs as the operational costs will be lower for a GSBR system compared to a conventional installation, unless you need big post- or pre-treatment facilities. The effect of suspended solids in influent and effluent will be studied during the pilot scale research. The outcome will decisive for the feasibility of the technology at full-scale.

Risks of implementing a new technology

Researchers working on technological innovation can not be familiar with all the details of all mechanisms involved. Innovations can have direct major consequences for humans and the natural environment, and thus technological innovation poses *potential* risks to its users and its environment. This argument also holds for aerobic granular sludge batch reactors.

In the paper, presented in session 7, the authors reported on the outcomes of a group decision room (GDR)-session where they asked specialists in the Netherlands, such as researchers, engineering firms, water boards and regulators about their opinion on which risks they thought to be relevant, the importance of these risks, and the phase of the research in which the risk had to be addressed. The authors asked the delegates of Granules 2004 workshop in Munich whether they agreed or disagreed with the outcomes reported at the meeting. During and after the discussion a number of interesting subjects were raised.

At the start of the discussion, two issues were raised concerning risks to public health. The first potential risk mentioned had to do with the mechanism of granulation, which possibly is related to hydrophobic interaction. If granulation goes hand in hand with favouring surface hydrophobicity and the composition of granular sludge consists of more hydrophobic bacteria than standard activated sludge, then granular sludge might contain more pathogens since, in general, pathogenicity and hydrophobicity are related. Additionally, when hydrophobic interaction between bacteria is an important condition for granulation, it might also be that the granule bacteria accumulate more easily in the human body.

The second point that was made concerned the risks of aerosols. If a granular batch reactor contains more hydrophobic bacteria than a standard activated sludge reactor, then the aeration in the bubble column may cause problems with aerosols. The hydrophobic material may concentrate on the borders between air and water, and the aerosols produced in the high reactor vessels may be more dangerous and spread over longer distances (at least one kilometre) than the aerosols of current activated sludge plants. One of the delegates conceded that when technological innovation is pursued, engineers run the risk of focusing primarily on the technological aspects, putting sanitary issues on a second place.

As the authors of the paper asked for the researchers' reaction to the GDR-session, it was suggested that lack of knowledge on part of the water boards was one of the reasons of the differences between the researchers and the water boards with respect to their estimate of the probability that aerobic granular sludge reactors will fail to meet the primary requirements. This was put in perspective by adding that it might also be a question of difference of responsibility. In the end, the water boards bear, at least formally, responsibility for the reliability of the plants, not the researchers.

Next, it was remarked that during the GDR-session, it was perhaps unclear what was meant by the term "pilot-plant". Did we mean that a pilot-plant had to be outside of the scientific laboratory or not? This ambiguity bears on the question, for instance, whether the use of raw municipal or industrial sewage water is allowed, or whether the pilot plant should use synthetic water. This possible uncertainty about the meaning of this concept may perhaps be one of the reasons for the voting outcomes underlying the discussion in the paper about the problem of many hands. (This question is answered in the paper mentioned above).

According to one of the delegates, the market potential for aerobic granular sludge batch reactors in Germany is low. There seems to be a lack of interest on the side of water boards due to conservatism. It was asked whether innovativeness is one of the benchmark requirements in Germany as it is the case in the Netherlands. The answer was negative. In the Netherlands, some water boards are willing to participate in innovative research since innovation is one of the five benchmark criteria for water boards.

The session ended with a vote by the delegates, as a group of scientists, on the question whether GSBRs would be able to eventually meet the primary requirements. Unanimously, it was believed that GSBRs would meet the primary effluent requirement provided that limited pre and after treatment is applied to remove suspended solids.

Where do we go from here?

During the different lectures and papers presented in the past and during the workshop, many different parameters of the granules were shown. However, de Kreuk found during reviewing the present literature that there was no clear structure in the parameters and methods that were reported. Many theories exist about the mechanism of granule formation, but no clear trends can be discovered when the used parameters and selection pressure is described in the papers. Therefore, a list of important parameters was discussed that should be included in literature about aerobic granule formation. The discussed parameters were the following:

- How (aerobic, anaerobic, dump-fill, plug flow, fluidised bed) and where (bottom/top of reactor, percentage of reactor height) is influent dosed and effluent extracted?
- What was the sampling procedure that was used? Since granules settle quickly, how was assured that a representative sample was taken.
- The applied oxygen concentration during the aeration period. Do not report values "larger than", but measure and report the oxygen concentrations.
- The selection pressure that is used to keep or form granules: applied superficial gas velocity; applied settling velocity (as in settling height/settling time), etc.
- Clear results towards granules formation, in order to indicate that the formed granules fit in the definition as given above and to indicate the stability of the system. So parameters that should be given are: average diameter, morphology, steady state time of the experiment, SRT, SVI_{10} and SVI_{30}. A desired parameter that gives a good indication of the granule morphology is the biomass density in grams of biomass per litre of biomass.
- Clear influent characteristics, not only the concentrations used, but also for example COD- and N-load. A discussion on the units in which the loads should be expressed did not lead to an agreement. To compare an aerobic granule system with conventional continuous systems the unit $kg \cdot m^{-3} \cdot d^{-1}$ would be suitable.
- Use recent references where possible.

As was shown during the workshop and in the different discussions, much research is already performed on the topic, but there are still unexplored fields in the world of aerobic granulation. These vary from pilot scale investigation to detailed research at microorganisms level. A brainstorm session resulted in a long list of research topics that are still not revealed. The topics considered:

- The granule as such (including mechanical strength of granules, is there a role for filaments in formation (spaghetti theory), what is the role and properties of EPS production, how is the mass transport inside the granule compared to mass transport in the surroundings, what is the exact role of shear force on the granule formation),
- The microbiological diversity of the granules (do filaments stay inside the granule or do they come in when circumstances change, what is the impact of protozoa (morphology), how do particles interact in or influence granule formation, is there a difference in sludge production/yield between granules and activated sludge, is horizontal gene transfer or modification of the biomass a method for improving reactor performance)
- Conversion processes that play a role (what is the impact of protozoa (kinetics), how are particles converted, how are polymers degraded and/or excreted, how does the structure

of the granules affect the conversion processes and how is mass transport involved, application of real wastewaters or complex substrates)
- General questions (as how are pathogens removed by aerobic granules compared to activated sludge and are there other risks and what are these risks of the use of aerobic granules).
-

The participants all agreed that general research on granule formation is ready to be brought to a larger scale of research: the pilot- or demonstration scale. Research topics that should be kept in mind at this scale are: stability of the granules and of the process and how could a reactor be (re)started, sludge handling/treatment, what are the characteristics of the effluent suspended solids and what type of post- or pre-treatment is needed, is the granule diameter controllable, what is the influence of mixing and how could this be done in large-scale reactors.

The conclusion that could be made from this session is that all participants agreed that up to now a lot of basic knowledge was gained on aerobic granule formation and that the pioneering stage is finished. Researchers should continue with specific research topics of which examples are given above and that the technology should be put into practice. Within a couple of years, this workshop should be repeated to see how we went from here; hopefully with new insights and with new applications.

References

Gjaltema, A., J. L. Vinke, M. C. M. Van Loosdrecht and J. J. Heijnen (1997). "Abrasion of suspended biofilm pellets in airlift reactors: Importance of shape, structure and particle concentrations." Biotechnol. Bioeng. 53: 88 - 99.

Schwarzenbeck, N., R. Erley and P. A. Wilderer (2004). "Aerobic granular sludge in an SBR-system treating wastewater rich in particulate matter." Wat. Sci. Technol. 49(11-12): 41-46.

IWA Publishing's authorised EU representative for General Product Safety Regulations is Diane D'Arras, 15 rue Duret, 75116 Paris, France, e-mail: safety@iwap.co.uk.

Printed and bound by CPI Group (UK) Ltd, Croydon, CR0 4YY
12/05/2026
02108863-0007